Route ONE Food RUN

Route ONE Food RUN

A ROLLICKING ROAD TRIP TO THE
Best Eateries
FROM CONNECTICUT TO MAINE
Plus Signature Recipes!

VINNIE PENN

Photography by
JENNIFER HIGHAM

Globe
Pequot

Guilford, Connecticut

Dedicated to Maredoll, for 100 years of greatness,
and bestowing upon me a love of books

All the information in this book is subject to change. We recommend that you call ahead to obtain current information before traveling.

Globe Pequot

An imprint of The Rowman & Littlefield Publishing Group, Inc.
4501 Forbes Blvd., Ste. 200
Lanham, MD 20706
www.rowman.com

British Library Cataloguing in Publication Information Available
Library of Congress Cataloging-in-Publication Data Available

ISBN 978-1-4930-2801-6 (paperback)
ISBN 978-1-4930-2802-3 (e-book)

♾™ The paper used in this publication meets the minimum requirements of American National Standard for Information Sciences—Permanence of Paper for Printed Library Materials, ANSI/NISO Z39.48-1992.

Printed in the United States

CONTENTS

CONNECTICUT

RHODE ISLAND

MASSACHUSETTS

MAINE

. . . And in the end, the food you make/ Is equal to the bread you break.

The first, and maybe only, time I ever had my heart broken was when I was seventeen years old. I recall not knowing what to do about the fact that I had zero interest in eating, and went an entire day without doing any. In retrospect, this is astonishing. I even made my favorite meal and sat on my front porch just staring at it, curious as to when this debilitating condition would subside. I confessed my situation to a cousin of mine, who said he knew exactly what I needed. This was initially disconcerting, as it reeked of ABC Afterschool Special territory, wherein he would follow up such a statement with a suggestion to do something absolutely crazy, and even illegal. I bit nonetheless, and his response was two words: "The road."

He had a penchant for melodrama and a VW bug with flames painted on the hood. His driver's license was still warm. We took to the open road, and damned if he hadn't nailed it. It came reeling at me, the white line and the wind, while I commandeered his radio to provide us with the appropriate soundtrack. I was feeling better already.

He begged off the highway after only a few exits, saying "This time of year it's all about Route One." Correct again. While its Rockwellian nature stung at first, this snapshot of the life I had envisioned with my senior-year girl-friend antagonized me, at once glorious and beyond reach, but before long the greenery and autumnal beauty rhapsodized. We bested the Rhode Island state line in record time, and the next thing I knew I was eating my first meal in over two days. The little log cabin–esque spot we descended upon is long since gone, its name eludes me, but I remember having the tastiest dish of fresh clam strips in my life, washing them down with a delightful Del's lemonade.

I solicited for fantastic spots along Route One, from my home state of Connecticut all the way to the tip of Maine, on both my morning radio show and social

media, and put together this roughshod list. We go off the grid a time or two, but it's well worth it.

Working on this book reminded me of that experience, of that time of my life. It also renewed my interest in Route One itself, a major artery in the northeast's chest.

Further, that old adage is obscenely right: The way to a man's heart is indeed through his stomach; it's just that no one ever talks about in a healing capacity. The food healed my heart that day, and the road my soul

— Vinnie Penn

A Note About Chefs

I've always regarded chefs as no different than artists. Why wouldn't I? The imagination, the eye for detail, the need for that most *particular* of touches, never mind the aesthetic end of the equation. But, where the artist's work can be admired year after year, gazed upon, discussed at length before guests, handed down from generation to generation, the chef's "thing of beauty" is typically engorged in the time it took for your table to place its collective order.

Poof.

Gone.

Maybe a photo taken to be posted on Instagram. (Probably.)

What I uncovered while researching this book, and meeting so many of the chefs involved—some eccentric, some not prone to small talk while others quite gregarious, and still others tripped up by humility while "talking shop"—was that not only was I correct, but chefs straight-up ARE artists.

Some of the culinary masters I encountered wrote their recipes out haphazardly on cocktail napkins in oftentimes illegible handwriting, others laboring over them afterwards, at their laptop, presumably as the sun rose, sleep-deprived and skittish. Some just began telling them to me, assuming it was 1977 and that I had a notebook and pen at the ready to take their account at the crime scene, seemingly struggling to even recall the dang thing, as if it's never made the same way twice. Not really anyway.

Even those who abhorred the word "chef," felt it not fitting, preferring cook, or even short order cook, simultaneously showing both a reverence and a disdain for pretension, had an artistic air about them, one going so far as to show me her recipe as it was originally written, by her later father, on a sheet of lined, yellowed paper, but not allowing me to even touch it, let alone "borrow it." I was instructed to take a picture of it with my cell phone, only to zoom in and transcribe it later in my office. Dad wasn't exactly a stickler for detail when it came to this dish either, easily the restaurant's most popular and an award-winner time and again.

The long and the short of it? All of those recipes, as I was given them, await you in this book. Some no more than three lines long, with others qualifying for novella status; some were written with a dinner party in mind, some for a family of four, or for two, or even just one, the guy or gal who put in a hell of a day and feels deserving of this at the end of it, despite being a table of one at home.

These are art in its purest form, nothing else, and my wish is that you'll see it that way, and be ready to play fast and loose in your kitchen while trying to bring one to life, even if many of these men and women creating these masterpieces in shotgun shacks up and down Route One loathe the word "fast" when it comes to food and would prefer "to taste" as opposed to "loose."

They have given us something precious, if not art then surely an invention, with many of these chefs/cooks/artists quite protective of what they were giving too, might I add, which—for some of them I suspect—might have lent itself to their perfunctory way of conveying said recipe (i.e. perhaps they'd prefer to keep some of it—even if just exactly how much paprika, or if there is any paprika at all—to themselves).

Take down the mixing bowl, pre-heat the oven, turn up the radio . . . your kitchen is a studio today.

— Vinnie Penn

CONNECTICUT

CAFÉ ALLEGRE

725 Boston Post Rd., Madison, CT
cafeallegre.com, (203) 245-7773

The late Chef Sylvio is kind of a big deal. Born Sylvio Suppa, he earned his Connecticut reputation in the kitchens of some of New Haven's finest Italian restaurants before taking his considerable skill-set to bucolic Madison. His café sits in the middle of the busiest part of town, across from the R.J. Julia Booksellers, and next to the Madison Arts Cinemas. In other words, you can spend your entire day in this one pristine spot.

SOUNDTRACK SUGGESTION

"Fix You" by Coldplay

Because what else does a fine meal do? Besides, the lush orchestral beauty of the British band's hit builds to such a sweeping, climactic finish, it is like a meal itself.

With several cookbooks of his own to his credit, Sylvio's specialty is Mediterranean cuisine, which he produces amongst the clatter and chaos of his kitchen without so much as breaking a sweat. After all, the slow cook is a thing; how else to make that pork shank fall apart with one gentle poke from a fork, to cut through some braciole effortlessly with that same fork? Sylvio knows this, revels in this, lives this.

His signature pumpkin risotto is a game-changer, and will make your dinner parties a must. Seriously—kids will befriend your children if only to score their mother and father an invite to your house once this gets around. For me, though, it's Sylvio's Osso Bucco Milanese, a breaded veal shank nestled onto even more risotto, as flavorful as the day is long, and as beautiful as the Madison night's sky.

Risotto with Pumpkin

SERVES 4

3 ounces (6 tablespoons) extra-virgin olive oil

2 cloves garlic, chopped

1½ pounds fresh pumpkin (or butternut squash), cut into large cubes

5 ounces dry white wine

Salt and pepper to taste

1 quart chicken or vegetable stock

1 package of risotto (4 ounces)

4 tablespoons grated Parmesan cheese

2 tablespoons chopped Italian parsley

Hot pepper (optional)

In a medium-size heavy saucepan, heat olive oil over medium-high heat. Add garlic and sauté, stirring continuously, until garlic is a light golden color.

Add the pumpkin (or squash), wine, salt, and pepper. Stir frequently until all liquid is absorbed. Add half the broth, cover, and simmer slowly for about 30 minutes, until pumpkin is tender.

While the pumpkin is cooking, prepare the risotto according to package directions (al dente is best). When both the pumpkin and risotto are fully cooked, mix the risotto into the pumpkin; if more broth is needed, add from the remaining stock until texture is creamy.

Top with parmesan cheese and parsley. Add hot pepper if desired.

HOME

1114 Main St., Branford, CT
homerestaurantct.com, (203) 483-5896

Smack dab in the middle of a bustling downtown area surrounding a quint-
essentially Connecticut green, Home is a Main Street mea culpa from what
can at times be an unforgiving road. Home is American cuisine 2.0, and owner/operator Jared Schulefand knows exactly what he is doing.

SOUNDTRACK SUGGESTION

**"Home Sweet Home"
by Motley Crue**

The decor may not lend itself to head-banging, but this heavy metal ode to all things home is easily the perfect song choice before dinner . . . and even after.

Even an order of fries is taken to the next level here, served with three different dipping sauces, and said sauces change every couple of months. In fact, Schulefand is keen on changing the entire menu every four months, overhauling it three times a year.

"We like to keep our customers on their toes," he says, with a mischievous glint in his eye. The "Home"-style mac & cheese might come mixed with meatballs or spicy buffalo chicken or even vegetarian enchiladas.

The Home Burger is topped with Boursin cheese, pecan-wood-smoked bacon, caramelized onion jam, and sautéed mushrooms. The certified Angus beef patty is served on a brioche bun.

If you sit up front you can grab one of the faux living room setups and share a couch with someone as you order fried dough chips, which are exactly what they sound like—tiny fried dough pizza chips, ready for dipping into a spicy marinara sauce. Or you can truly toast the open road with a cold glass of Sauvignon Blanc and the Swordfish Saltimbocca. Divine. Dine.

Dinner and a Movie

Dinner and a movie constitutes the quintessential date night, but I refuse to play a part in assisting your significant other, or running buddy, or bestie, in simply traipsing to the closest mega-cineplex after dinner at one of these ridiculously cool spots. Fact is, along Route One, or in the vicinity thereof, there are a litany of gorgeous old movie houses with history running through their wooden veins, most dedicated to arthouse fare, with but a few exceptions.

Madison Art Cinemas
761 Boston Post Rd., Madison, CT
madisonartcinemas.com, (203) 245-3456

At over 100 years old (it opened in 1912), this is one "oldie movie house" that eschews novelty spins of classics on rainy Saturdays in the fall in favor of first-run indie and foreign films. You'll find no midnight showings of The Rocky Horror Picture Show here. You will find a little cafe set aside from concessions should you need a little espresso before the latest tiny budget, British Oscar hopeful. The decor is charming, and there are only two films running at a given time. The Madison Art Cinemas doesn't cram its history down your throat, but it's there just the same.

WEST END BISTRO

551 Main St., West Haven, CT
westendbistro.net, (203) 691-9223

There are no late-night bites here. This relative newcomer is all about being the place you plan on going to, not the one you wind up at. The latest they are open is 8 p.m.

They serve breakfast all day, and one would be hard-pressed to stray from that menu. Have them scramble up a couple of farm-fresh eggs and serve them alongside their pork belly grits loaded with roasted garlic and cheese. Or, similarly, order a bacon-wrapped hot dog and have that served with an order of West End madness—truffle parmesan home fries with grilled cremini mushrooms.

SOUNDTRACK SUGGESTION

"West End Girls" by Pet Shop Boys

The play is in the title, of course, when it comes to this '80's classic, but the jaunty melody fits perfectly with this joint's atmosphere, even if West Haven and London have zero in common otherwise.

Truly, with a breakfast menu boasting items such as "cutlets and flapjacks," which is just that, two thick, fluffy pancakes and two chicken cutlets, spiked with a maple-jalapeno dulce de leche sauce, an actual lunch menu, let alone dinner, seems pointless.

That said, there are vegan street tacos that'll knock your socks off, with house-pickled veggies spilling out of the shells, or a—get this—cheeseburger salad. It's a menu so innovative, so outrageous, you just can't miss. Chef Jeff Lamberti is a madman. Viva la West End!

Crab Cake Imperial

SERVES 4

FOR THE CRAB CAKES

1 teaspoon oil

2 eggs

½ cup mayo

1 tablespoon Worcestershire sauce

1 tablespoon lemon juice

1 tablespoon Dijon mustard

1 green pepper, diced and cooked until soft

1 Spanish onion, diced and cooked until soft

½ teaspoon salt

1 to 2 teaspoons black pepper

¼ cup chopped cilantro

¼ cup chopped chives

3 cups panko

1 tablespoon Old Bay Seasoning

1 pound lump crab meat

FOR THE IMPERIAL SAUCE

1 tablespoon oil

¼ cup chopped shallots

1 tablespoon chopped garlic

1 cup white wine

1 cup seafood stock

8 tablespoons butter

1 bunch of fresh thyme

Mix oil, eggs, mayonnaise, Worcestershire sauce, lemon juice, and mustard together.

Incorporate green pepper, onion, salt, black pepper, cilantro, chives, panko, and Old Bay Seasoning.

Fold in the crab, being careful to not break up the crab meat too much.

Portion into patties and chill for a minimum of two hours).

Heat oven to 350°F.

In an oven-safe pan over medium heat, heat oil and brown crab cakes on both sides, then remove from pan.

Add shallots, and sauté until slightly brown.

Add garlic, wine, and stock and return crab cakes to pan. Add butter and thyme.

Bake for 5 to 10 minutes until sauce reduces. Garlic and shallots cook in heat left over from patties, and when patties are returned to pan, cover, and continue cooking until reduction occurs.) *A watchful eye leads to a FULL belly too!

SEVEN SEAS RESTAURANT

16 New Haven Ave., Milford, CT
7seasmilford.com, (203) 877-7327

This is the place that makes it feel like summertime even while a mean winter wind blows outside. With a side order of flurries and excruciatingly cold temps, it's always summer at 7 Seas.

If you want hot, spicy wings, they have you covered. Ditto for burgers and all the usual suspects appetizer-wise, but the seafood rolls are where it's at. Shrimp, clam, lobster . . . heck, if you can't decide you can get the combination roll, which enables the guy in the kitchen to load your buttered roll with a few choice strips of each.

The Peachtree Chicken is a standout on the dinner menu, where some awfully tender chicken breast is sautéed with peaches and walnuts and then glazed with peachtree schnapps, which takes me back to roller rinks in 1980. Somehow it works quite well.

When you need to hide from the elements, head to 7 Seas where you can hide from the world even, but where you can easily be corralled as well.

SOUNDTRACK SUGGESTION

"Come and Get It" by Badfinger

Written by Paul McCartney himself, what says TLC more than a Beatle giving up a tune for the greater good? And despite what the Fab Four were becoming known for, it was as simple as that: If you want it, come and get it. You should. Good food demands that you do.

Fish 'n' Chips

SERVES 2

2 large eggs

Salt and pepper to taste

6 4-ounce filets of cod or haddock, patted dry with paper towel

1 cup flour

2 cups extra-fine cracker meal

Paprika to taste

Sea salt to taste

½ cup light canola oil

In a small bowl, whisk the eggs, adding a pinch each of salt and pepper.

In a separate bowl, place the flour.

Roll the fish fillets in flour, shaking gently so the excess falls off, then dip floured fish in egg.

Mix paprika and sea salt with the cracker meal.

Lightly bread the fish with cracker meal mixture.

In a skillet, heat canola oil to high heat. Place breaded fish into the hot oil. Cook for 2 minutes on each side, until browned. Place on paper towel–lined plate to soak up excess oils.

Add more paprika and sea salt if desired.

CHEF'S TIP

To avoid the greasiness and heaviness that usually happens when you make fish and chips, deep fry only in light canola oil. Also, do not use breadcrumbs to bread the fish. "That is how the fish becomes heavy and inevitably greasy, sang a chorus of enthusiastic kitchen workers, from the sous chef to the executive one, and maybe even a bar-back. The love was palpable.

GOODIES

111 Boston Post Rd., Orange
goodiesorange.com, (203) 553-9944
(Other locations: 170 Cherry St., Milford,
and (seasonally) on Gulf Beach in Milford)

Goodies hammered an exquisite nail in the coffin of my "I'll never eat a gyro" crusade. Born during my college years, when a female friend bit into one and the pita evacuated its contents all over her hoodie, this crusade looked to be taken with me to the grave. What I thought looked like Thousand Island dressing leapt across my friend's hoodie logo with terrifying ambition, as if no pita could ever contain it. It looked repugnant. It was the stuff of dry heaves.

But Yanni at Goodies insisted I give a Goodies gyro a whirl. What's more, he informed me that a true Greek gyro would have tzatziki sauce enveloping the lamb, beef, and pork, rhapsodizing the onion, lettuce, and cucumber. The clouds parted this day, as they can for you, while I noshed on the fluffiest of pitas and enjoyed my first gyro. It surely won't be my last.

Goodies is what many road trippers consider to be the perfect pit stop. It's a breakfast, lunch, and dinner spot, and a quick one at that, with numbers being barked out rapid-fire. "38! Serving 38! 39! I'm now looking for 39 too!"

The Goodies breakfast consists of two sizable pancakes, with a fried egg nestled into each respective circle, appearing as if wide eyes are gazing back at you. Delicious home fries accompany these two breakfast staples. You'd be hard-pressed to not jump right into your car and start your day after this breakfast feast.

As for traditional drive-in fare, Goodies has you covered. A burger all their own, with onion rings placed right on top of the patty rather than on the side, rivals all Five of the Guys, and the clam strips are juicy and tender and render the guy with the spatula shouting, "They're not frozen!" completely unnecessary.

SOUNDTRACK SUGGESTION

"Little Deuce Coupe" by The Beach Boys

From the classic cars that often grace the parking lot (there are classic car shows that actually take place in the lot of the Milford location), to that '50s and '60s drive-in feel, any ol' Beach Boys song will do, but this is nail-on-the-head territory.

Over Easy

SERVES 1

2 large eggs

1 tablespoon butter

1 pinch kosher salt

1 pinch Black pepper

Crack eggs into a bowl. This is your best chance at getting them to cook evenly, plus you have more control over placement into pan.

Heat skillet over low heat and add butter. As it melts, brush the butter around the pan.

When the butter stops foaming, pour the eggs into the pan, then quickly lift the handle just enough for the eggs to pool slightly on the far side. This will prevent running. After 10 to 15 seconds, smoothly lower the handle. Wait another 10 seconds, then lightly jiggle the pan just to make sure nothing is sticking.

Season with salt and pepper and cook over low heat, for 1 to 1½ minutes.

Remove pan from stove top and either attempt a pancake-style flip (difficult), bringing the pan UP to the egg, or flip briskly with spatula.

Return the pan to low heat and slowly count to 10. Re-flip the eggs to their original side. (This time it won't be so difficult.)

Slide onto a warmed plate and serve immediately with toast for wiping up all the orangey delightfulness.

JOHNNY AD'S

910 Boston Post Rd., Old Saybrook, CT
johnnyads.com, (860) 388-4032

Johnny Ad's has the look of one of those places you just know is good as you're buzzing by it with the top down on a spring afternoon. To call it a shotgun shack is fitting; it can claim you from the road—just beg ya to throw the car in park, if only for some takeout. It simply reeks of foot-long hot dogs, out-of-this-world shakes, and—hell yeah—lobster salad. (Recipe below).

A Saybrook staple, Johnny Ad's can make it feel like summer in the dead of a New England winter. The top notch, tried-and-true summer favorites-style menu has the ability to transport, to transcend. The only thing missing is a beach blanket for you to be able to throw on the floor to dine on; and there's gotta be health regulations there.

Small and unassuming, the portions, interestingly, are the exact opposite: sizable, and assuming you're starved. The foot-long hotdog somehow appears to be at least a foot and a ½, and the not-too deep-fried onion rings are akin to the circle of moisture your glass of lemonade would leave if not placed on a coaster, plus there's no way you're not taking some of them home with you. So, if they're not gonna give ya the beach blanket, they may as well just bring the doggy-bag to the table with your meal.

SOUNDTRACK SUGGESTION

"Take the Long Way Home" by Supertramp

The town of Old Saybrook is worthy of tooling around, thanks to its proximity to the water and vintage New England charm. This early-'80s pop anthem offers sage advice indeed.

Lobster Salad

MAKES 3 TO 4 LOBSTER ROLLS

1 pound lobster meat, cooked and cut into ½-inch chunks

1 pinch celery salt

Juice of ½ lemon, freshly squeezed

Mayonnaise to taste

3 or 4 split top rolls

Chopped mixed greens (think aesthetic versus appetite)

1 large tomato, sliced thin (optional)

In a medium bowl, mix lobster, celery salt, and lemon juice. Stir in the desired amount of mayonnaise. Place in refrigerator to chill while you prepare the rolls.

Lightly toast the rolls and line with mixed greens and tomato if desired. Top with chilled lobster salad mixture.

JAKE'S DIGGITY DOGS

216 Crown St., New Haven, CT
jakesdiggitydogs.com, (203) 782-1111
(Two more locations are in Cromwell and Branford.)

Jake is pretty much up for putting anything on top of a hot dog. This is that place. Macaroni and cheese, cream cheese and bacon, pastrami and swiss (you guessed it—the Reuben) . . . you name it, they'll do it. Crammed between two thumping clubs in the thriving downtown New Haven theater district, Jake's is a joint loaded with character, run by a character, and serving plenty of characters at the same time.

Jake—last name Russell—is a living, breathing New Haven landmark himself. A key figure in the Nighthawks organization of the late '70s, early '80s, Jake has framed photographs of hockey players all over the walls of the tiny establishment that seats no more than a dozen people at best. There's also a framed photo of Jim Morrison being busted downtown after a legendary Doors gig at the now-defunct New Haven Coliseum. Like I said, it's that place.

He puts the baked beans directly on top of the dog, as opposed to alongside it, and does the same with the coleslaw. "The Hawaiian" is smothered in pineapple chunks and coconut shavings.

Speaking of Hawaii, and as if all this decadence wasn't enough, Jake also serves twenty-one flavors of mochi ice cream. "The mochi was already here," he'll tell you, downplaying its innovative inclusion on such a decidedly Americana menu. It's actually Japanese by way of Hawaii and is basically pounded sticky rice with ice cream inside. Flavors include vanilla, chocolate, strawberry, Kona coffee, plum wine, and red bean. It takes a purveyor

SOUNDTRACK SUGGESTION

"Hot Child in the City" by Nick Gilder

This classic '70s ditty actually sizzles, in my humble opinion, and reeks of the hot child bobbing and weaving through the bustling streets of his hometown, hot dog in hand, two bites deep.

of delicacy to see these two worlds (hot dogs and mochi) colliding, and Jake did. He's that kind of guy.

Jake's Cole Slaw
SERVES 6

½ cup mayonnaise

⅓ cup milk

1 teaspoon white vinegar

¼ cup sugar

¼ teaspoon salt

1 (16-ounce) package cabbage coleslaw mix (see Options)

In a large bowl whisk together all ingredients except coleslaw mix until smooth and creamy.

Add coleslaw mix and toss until well coated. Cover and chill for at least 1 hour before serving.

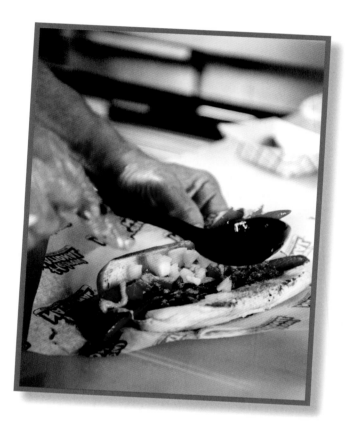

CRISTY'S MADISON

73 W. Wharf Rd., Madison, CT
cristysmadison.com, (203) 245-7377
(Second location at 1261 Boston Post Rd., Westbrook)

Cristy's Madison is ultimately all that's left standing where the once-great Dolly Madison stood, a legendary Connecticut inn kneeling on Madison's shores. The lights may have dimmed on that towering establishment, but the crowd still comes for Cristy's, and why wouldn't they? There are still rooms to be had, views to be taken in, and now a menu many on the shoreline were already familiar with thanks to the Cristy's "up the road" in Westbrook. "Up the road" is a gentle euphemism for the foodie/road tripper that can actually translate to "an hour away" for the neophyte foodie. In this case it's maybe twenty-five minutes, depending on traffic.

SOUNDTRACK SUGGESTION

"Thunder Road" by Bruce Springsteen

With The Boss getting all plaintive on us, waxing nostalgic with feet firmly in the present at the same time, Cristy's feels as if it could totally be a waterside Jersey haunt from Bruce's early years.

Cristy's looks and feels retro—heck, it can even taste retro—as boats wobble on trailer hitches either toward the water or away from it depending on the time of year. People walk golden retrievers or careen by on roller skates as you gaze out the window, listening to the aforementioned waves make their music.

But, let's talk pancakes. Cristy's is probably best known for them, and does do the breakfast boogie all day. But they also manage to offer tried-and-true American fare (insane meatloaf, even insaner BBLT, a bacon, lettuce, and tomato sandwich with extra bacon, served on Texas toast to boot) and a hot dog selection that rivals the pancake one. The Coney Island dog might be without peer, which is less hubris than it is satisfied customer. The chili is a meal unto itself, not running down your hand or making a break for your plate with every bite. But it's the Westbrook dog, armed with onion rings and cheddar cheese, that is truly ready for battle. The stuff of road trip fuel.

Lobster Benedict

SERVES 4

FOR HOLLANDAISE

4 egg yolks

3½ tablespoons lemon juice

1 pinch ground white pepper

⅛ teaspoon Worcestershire sauce

1 tablespoon water

1 cup butter, melted

¼ teaspoon salt

FOR POACHED EGGS

1 teaspoon distilled white vinegar

8 eggs

Picked lobster meat - working from 4-6 lobster tail halves

8 strips Canadian-style bacon

4 English muffins, split

2 tablespoons butter, softened

1 tablespoon chopped chives

For Hollandaise

Fill the bottom of a double boiler part way with water, and bring to simmer.

In the top of the double boiler, whisk together egg yolks, lemon juice, white pepper, Worcestershire sauce, and water.

Add the melted butter to egg yolk mixture 2 tablespoons at a time while whisking yolks constantly. If it gets too thick, add a teaspoon hot water. Whisk in salt and remove from heat. Cover and keep warm.

For poached eggs

Bring 3 inches of water to a simmer in a large saucepan, then add vinegar. Carefully break eggs into simmering water, allow to cook for 2 to 3 minutes. Remove eggs from water with slotted spoon, and set on warm plate. Add picked lobster meat to hot water to heat.

Meanwhile, brown bacon in medium skillet over medium-high heat and toast muffins.

Spread softened butter on the muffins, top with cooked bacon, followed by one poached egg per. Drizzle with Hollandaise. Sprinkle with chives. Serve immediately.

BISTRO MEDITERRANEAN TAPAS BAR

383 Main St., East Haven, CT
bistromediterraneanandtapasbar.com, (203) 467-2500
(There are also locations on Route One in Norwalk and Westbrook)

Bistro Mediterranean Tapas Bar—known simply as Bistro by the locals—manages to pull off a feat like no other. Its surroundings are pretty much completely devoid of scenery, yet Bistro transports its diners to an island before they are even seated. Actually, make that the Spanish countryside. Once you are gorging on Bistro's exquisite Spanish peppers, I defy you to hear the hustle and bustle of the traffic just outside the door, the hum of Main Street.

SOUNDTRACK SUGGESTION

"Whatever, Whenever" by Shakira

This song captures Bistro's spirit handily—the menu begs the diner to order whatever, whenever, and the rhythm of the Shakira star-maker dials in to the restaurant's motif.

Let's discuss these Spanish peppers in detail, shall we? On the menu as "Piquillos Rellenos de Cordero," they are stuffed with braised lamb and served with a port wine and shallots sauce, and are flavorful enough to make your lunch hour feel like five of 'em. The work you left behind will be met with renewed vigor upon your return.

Interestingly, one of the most talked-about dishes at Bistro is the epitome of a tapas. It's a Brussels sprouts appetizer. I'm not kidding. Brussels Salteadas is the go-to dish at Bistro. Sautéed with garlic, olive oil, white beans, and Serrano ham, the Brussels sprouts simply come to life, and become as "poppable" as a jalapeno popper, yet significantly better for you.

The Ensalada de Alcachofas pairs artichoke hearts with the spicy pepper, cilantro, olive oil, lemon juice, and ricotta salata cheese. This is fusion at its finest, kicking an artichoke appetizer up considerably, and leaving you more than satisfied. In fact, you'll feel as if you've vacationed.

Shaved Brussel Sprout Salad

MAKES 4 SERVINGS

FOR THE SALAD

1 pound Brussels sprouts, shaved using a mandolin

¼ cup black truffles, chopped

3 tablespoons black truffle oil

¼ cup grated Parmesan cheese

Juice of 2 lemons

FOR THE RISOTTO

½ tablespoon oil for frying

½ cup diced yellow onion

1 cup Arborio rice

2 cups water

Salt and pepper to taste

½ cup butter

½ cup cream cheese

⅓ cup Parmesan cheese.

All-purpose flour for dusting

For the salad

Mix all the ingredients in a bowl, salt to taste and chill.

For the risotto

Sauté onions in a deep pan until soft. Add rice, water, salt, and pepper, and boil on high for 10 minutes.

Reduce to low heat, cover, and cook for another 10 minutes until rice is firm but not mushy.

Transfer to a mixing bowl and while it is hot, add butter, cream cheese and Parmesan, mixing until it is creamy and smooth.

Transfer mixture to an oiled, 1-inch-deep sheet pan. Make sure the mixture forms a layer ¾ of an inch thick while pressing and tightening. Cool down and refrigerate until hard.

Heat a deep fryer to 350°F. Cut risotto into 3-inch-square pieces, dust with flour, and fry until golden.

Serve chilled salad over risotto.

THE RIB HOUSE

190 Main St, East Haven, CT
theribhouseeh.com, (203) 469-7427

Recently relocated from its longtime location at what locals call the Branford hill, The Rib House is now a full-on East Haven institution. Long before "BBQ" became a battle cry for restaurant chains and New England weekend warriors in the summer months, this place was serving up the most mouth-watering, fall-off-the-bone bad boys around. Suck-the-sauce-off-your-fingers good is what I'm saying.

But you can stray. From the ribs, I mean. The burgers are epic, and oftentimes necessitate a doggy bag. The nachos are out of this world, and I've never quite seen them presented the way they are here . . . ever. Like butter-flied nacho shells, the chips are sizable and smothered in Monterey Jack. Truly a meal unto itself.

The Rib House is a "best kept secret" kind of place, where regulars are the bread and butter, and that's just fine. The phone number for the place is 469-RIBS, and ravenous customers blindly call year after year, at a maddening rate, inquiring about the ridiculously reasonable, to-die-for $4.69 rib meal.

SOUNDTRACK SUGGESTION

"867-5309"
by Tommy Tutone

I mean, the *number's on the wall.* Okay, so maybe that's not enough for this selection. How 'bout, then, the fact that the repeated "I got it, I got it" works perfectly as far as ribs go, and holding one up while singing along to this one-hit wonder's ode to a girl he probably never even had the nerve to call?

Mona's Cheesecake

MAKES ONE CHEESECAKE

FOR THE CRUST

2 tablespoons granulated sugar

½ stick melted butter

1½ cups crushed graham crackers

FOR THE CHEESECAKE

4 (8-ounce) packages cream cheese, room temperature

2 (16-ounce) packages sour cream

6 eggs

2 cups granulated sugar

2 teaspoons vanilla

For the crust

Mix sugar and butter with graham crackers. Pat down into a 10-inch cheesecake pan.

For the chesecake

Heat oven to 350°F.

In a large bowl, beat the cream cheese and sour cream together.

Beat in eggs, one at a time.

Mix in the sugar and add in vanilla.

Pour mixture on top of crust and bake for 55 minutes. When the timer goes off, keep oven door shut and let cake sit with the oven off for another 55 minutes.

Remove and let cool for 2 hours. Refrigerate and enjoy!

BIN 100

100 Lansdale Ave., Milford, CT
bin100.com, (203) 882-1400

Binn 100's Chef Elena Fusco comes from a family of chefs. Her brothers own other notable New Haven eateries, even more off the beaten Route One path but equally worthy of a visit. Elena's may be closest to our locale, and is incredibly reasonably priced Italian/Mediterranean fare that will more than sustain you on your road trip. Plus, it's elegant enough to pop the question, and subsequent cork.

SOUNDTRACK SUGGESTION

Billy Joel's "Scenes from an Italian Restaurant"

The fact is, this track works with many of the restaurants in this book, but Bin's enormous wine selection and traditional Italian dishes especially evoke the Joel classic.

But the kids route works too, which is the beauty of the Bin. The nostalgic (that's you, mom and dad), will enjoy one—if not all—of the Little Rascal Pizzas. The Spanky packs a considerable punch via chipotle chicken, gorgonzola, and caramelized onions, while The Buckwheat boasts sausage, broccoli rabe, and mozzarella. The Alfalfa is my particular favorite, topped with goat cheese, roasted red peppers, mozzarella, prosciutto, and arugula. You get extra points if you say you're in the mood for it because you're the Barber of Seville.

The Bin 100 staple is pan-seared scallops served over homemade lemon pepper fettuccine in a pureed sofrito sauce of garlic, bell peppers, onions, and fresh tomato, and finished off with touch of cream. The scallops are juicy and spiced to perfection.

Elegant and cozy all at the same time, the award-winning Bin 100 makes the cut because sometimes the road is so demanding you need to feel as if you're celebrating the achievement of a family member, and the menu and atmosphere here can definitely assist in fooling yourself. Sometimes, that's an achievement in and of itself.

Shrimp and Scallop with Creamy Lobster Sauce

SERVES 4

2 eggs

Salt and pepper to taste

1 cup flour

1 pound large sea scallops, patted dry with paper towels

½ cup vegetable oil

1 pound extra-large shrimp, peeled and deveined

Olive oil for drizzling

1 teaspoon butter

1 fresh shallot, finely chopped

1 teaspoon lobster base (preferably Knorr)

1½ cups heavy cream

Preheat the oven to 400°F.

In a small bowl, whisk the two eggs, adding a pinch of salt and pepper. Set aside.

In a separate bowl, add flour and dredge the scallops through it, shaking off excess so they are lightly dusted.

Heat vegetable oil in a 10-inch frying pan to high heat. Working swiftly in small batches dip the scallops in the egg, allowing any excess to drain and place into the hot oil. Cook 1 to 2 minutes on each side until browned. Remove from frying pan and transfer onto one half of the baking sheet.

Arrange the raw shrimp on the remaining side of the baking sheet and sprinkle with salt and pepper and a small drizzle of olive oil.

Place the baking sheet on the center rack of the oven and cook for 10 minutes until the shrimp are pink and opaque.

While the shrimp and scallops are cooking, prepare the sauce on stovetop.

In a 1-quart pot over medium high heat add butter and shallot and sauté for 2 minutes or until shallots are translucent but not browned.

Add the lobster base and heavy cream to pot. Raise to high heat and bring to a boil while whisking the mixture. Reduce to medium heat and cook while whisking until the sauce is reduced to half the volume and thickened. Add a dash of pepper and taste for saltiness, if needed you can add a pinch of salt to taste. Remove from heat and let rest.

Transfer the seafood to a platter and spoon the sauce over the top. We like to serve this dish with sautéed spinach and mashed potatoes.

_____

CHEF'S TIP
We recommend using the Knorr brand of lobster base. It can be ordered on Amazon and has a one-year shelf life.

STOWE'S

347 Beach St., West Haven, CT
stowesseafood.com, (203) 934-1991

You'd be hard-pressed to career through West Haven while on a Route One Food Run and not make a quick pit-stop at the beach. This is where Stowe's comes in to play. With the overturned rowboat out front, and myriad sea-life accessories adorning the door and windows, it can be both a break from the breakwaters of Long Island Sound and the traffic jams that force folks onto Route One in the first place.

SOUNDTRACK SUGGESTION

"No, Shirt, No Shoes, No Problem"
by Kenny Chesney

Stowe's has the no-shirt, no-shoes, no-problem thing down as much as any of the places from Rhode Island we've included here, but the owner and the staff give off the most easy-going air ever, plus you can hear water lapping upon entry.

Stowe's goes the distance in the condiment category, which is refreshing. Any joint willing to churning out their own tartar and cocktail sauce (recipes below) isn't going to dump said sauces on microwaved morsels. Stowe's is the epitome of fresh, with fresh air in your face to boot.

I'm a sucker for their seafood chili, a heartening bowl of hot chili where the ground beef is bumped in favor of savory sole, scallops, and shrimp. Speaking of shrimp, from the cocktail to the fried roll, you're in good hands—and they'll be in yours only briefly. Stowe's knows what it is, and there is zero pretense. It's a pirates' refuge at which you are meant to stop down for but a moment.

Tartar Sauce

1 cup heavy mayonnaise

¼ cup sweet pickle relish

1 tablespoon capers

1 small onion

In a small bowl mix the mayonnaise with the sweet pickle relish. Mince the capers and the onion and add to the mayonnaise mix. Mix well and refrigerate for at least 1 hour prior to serving.

Cocktail Sauce

1 cup ketchup

⅓ cup ground horseradish

½ teaspoon hot sauce

Thoroughly stir together all ingredients and chill before serving.

DIVE BAR

24 Ocean Ave., West Haven, CT
divebarandrestaurant.com, (203) 933-3483

If it is later in the day, and a sturdy cocktail is as much a must as a meal, might I suggest the glorious, notorious Dive Bar. The place gives new meaning to the term watering hole.

SOUNDTRACK SUGGESTION

"When the Sun Goes Down"
by Kenny Chesney

This entry is the perfect companion piece to the aforementioned—and I mean both the songs and the establishments. By all means head to Dive Bar when the sun goes down, but after a good, long day on the beach and lunch at Stowe's. It's where the locals turn it up, and this is where Chesney does too, with Uncle Kracker right alongside him!

Now, I don't know about you, but I'm a fan of this taco renaissance. I say, put anything and everything into one of those crunchy shells. I vividly recall the day I saw one filled with ice cream and coated in chocolate and said to the Good Humor Man, "This is right. What took so long?" I was eight.

Dive Bar clearly agrees, as their Mahi Fish Taco (recipe below) belongs on the Sunset Strip. It's collision of cilantro and smoked paprika takes it to the next level and renders it a dream when washed down with a Mojito.

Dive Bar has you covered craft beer wise too. From Ithaca to Brooklyn Brewery, they've got the Northeast corridor sewn up, but don't come close to stopping there. They've pretty much got a burger to go with every beer, from the New England (glazed apple, bacon, and bleu cheese) to the obscene French Pig (ham, pulled pork, mushrooms, and swiss), but this place couldn't possibly call itself Dive Bar and not have a burger topped with a fried egg. Yes, "The Hangover" is on the menu. Dig the irony.

Mahi Fish Tacos

SERVES 3

FOR THE SALSA

1 tomato

1 mango

1 red onion

1 pineapple, cut in quarters

Pinch salt

Pinch fresh ground pepper

2 tablespoons sugar

3 tablespoons cilantro

Zest of 2 limes

1 jalapeno pepper, seeded and chopped small

FOR THE MAHI

6 ounces fresh mahi, skin removed

Salt and pepper to taste

1 tablespoon smoked paprika

½ tablespoon cayenne pepper

1 lime wedge per fillet

1 head romaine lettuce, shredded

3 6-inch flour tortilla shells

For the salsa

Dice tomato, mango, half red onion, and quarter of pineapple (reserve other 3 quarters).

Place in mixing bowl, and add salt, pepper, sugar, and cilantro and mix.

Squeeze lime zest over salsa and add jalapeno to taste.

For the mahi

Heat oven to 375°F.

Sprinkle fish with salt and pepper.

In a small bowl, mix together smoked paprika and cayenne pepper and sprinkle on top of fish.

Place fish into hot, oven-safe sauté pan and cook for 3 minutes on each side. Once fish is browned, squeeze lime wedge over fish, and then place wedge directly on the fillet. Keep mahi in pan and place in oven until fish becomes tender and starts to flake. Remove lime wedges.

Place shredded lettuce in taco shells. Spoon fish in on top of the lettuce and top with salsa. Enjoy! Of course, tacos are a matter of preference - perhaps you prefer no lettuce. Same with the shell - warm if you'd like, or leave at room temperature for a more corn chip taste. Add anything else you might like, from black olives to diced onions. Get assembly line, baby.

Dinner and a Movie

Once you've set your sights on New York City, should you be passing through Connecticut from, say, Rhode Island or beyond, it can get pretty easy to think, "In just few miles more there will be great food and fabulous little theaters." Yeah, don't think like that. Don't sell this small state short, and don't miss out on Garden Cinemas.

Garden Cinemas
26 Isaacs St., Norwalk, CT
(203) 838-4504

The beauty of Garden Cinemas' proximity to The Big Apple is that movie distributors many times view arthouse cinemas within driving distance of the city as a practical place to try out the little movie that could. "The Garden," as it is known, is also keen on urban features, those films bound to succeed in Harlem, Queens, and the like, that could easily take off at a venue like this, which is just off our beaten path.

SCARPELLINO'S

257 Forbes Ave., New Haven, CT
scarpsnewhaven.com, (203) 468-7827

Potatoes and eggs, hot dogs in tomato sauce, and suffrito are the staples of an Italian kitchen, and every last one of 'em comes out of Scarpellino's, with each meal potentially reminding you of your childhood as well.

SOUNDTRACK SUGGESTION

"My Way"
by Frank Sinatra

It'd be an insult to listen to anyone other than Sinatra when it comes to Scarpellino's, such fans are they, but the song works in that you can certainly order things your way, too!

Scarp has attached some fun names to the flavorful fare. For instance, the suffrito can be found on the considerable menu labeled as "That Old Devil Moon," and the potatoes and eggs are the "Nice 'n Easy." There's broccoli rabe here, too, and it is quite possibly without peer.

Scarpellino's is an institution, having moved around New Haven a few times since its debut in 1981. Nothing but Sinatra comes through the speakers, and the decor is, I'd estimate, 80 percent Sinatra too.

The Penne Francesca is as bountiful as it is beautiful, with the penne pasta awash in Scarpellino's rich vodka cream sauce, and accompanied by tasty chicken sausage. You are transported, I tell ya, to a place where the guys and dolls are wise and molls.

Chicken Dominic
SERVES 6

¼ cup olive oil

1 onion, chopped fine

4 cloves garlic, chopped fine

¼ teaspoon hot pepper flakes (optional)

6-8 boneless chicken (preferably tenders), rinsed and patted dry

1 cup roasted red peppers

½ cup crumbled gorgonzola cheese (not dressing)

8 ounces pasta of your choice (rigatoni is a great choice)

In a sauce pan, heat olive oil to medium heat. Place onion, garlic, and hot pepper in oil and brown lightly.

Reduce heat to low, and place chicken in pan and cook for 10–12 minutes. Add red peppers and gorgonzola. Cheese will melt quickly, creating sauce.

Cook pasta according to package instructions. Strain.

Toss pasta together with cheese sauce and serve immediately.

TURTLE CAFE

4 Westbrook Place, Westbrook, CT
facebook.com/theturtlecafe, (860) 399-1799

The Turtle Cafe is a quick detour off Route One, less than a minute from the hustle and bustle. The menu is tirelessly imaginative, with daily specials that surprise as much as they entice. If you're lucky you'll get a table, or you can wind up at the counter, perhaps even next to a regular whose name graces the menu. Rumor has it one regular has four dishes named in his honor!

SOUNDTRACK SUGGESTION

"Buddy Holly"
by Weezer

Sure, you can go for an actual Buddy Holly tune if you're so inclined, but since the proprietors have gone the distance to make it feel like yesterday for you, why not let alt rockers Weezer do the same with this poppy, perfect gem from the mid-'90s?

You'll almost feel transported to another time and place. The Mayberry/yesteryear feel of the entire block is a welcome distraction from the road, what with an actual general store just doors down.

It appears as if crepes and/or waffles are the must here. Loaded with every fruit imaginable, and topped with a raspberry sauce and whipped cream, they put my photographer in a crepes coma after her shoot. It is the definition of rich. And filling.

Turtle Frittata

SERVES 2

4 eggs

¼ cup milk

butter, margarine, or olive oil to coat pan

1 small white onion, or half a large onion chopped lengthwise.

½ cup spinach

4 ounces goat cheese, crumbled

1 tablespoon Italian herbs

Mixed greens, optional

Preheat oven to 350°F.

Whisk eggs with a splash of milk and set aside.

Melt butter/margarine/olive oil in a large, oven-proof skillet. Add onion and sauté until brown, stirring occasionally.

Add spinach and the whisked eggs.

Add crumbled goat cheese and Italian herbs, and place skillet in oven for 15 minutes or until frittata browns.

Top with mixed greens or eat as is. Enjoy!

CHEF'S TIP

This dish is fully loaded with protein and a very healthy way to start the day. Have it for lunch or brunch!

PARTHENON RESTAURANT

809 Boston Post Rd., Old Saybrook, CT
parthenondiner.com, (860) 395-5111
(Additional location at 374 East Main St., Branford, CT)

Remember when you were a kid and your mom threw you a curve ball at dinner time and plunked pancakes down in front of you? Or a waffle hot off the waffle iron? And you were like, "What's going on? Is there a hidden camera? Why are you serving breakfast at night?" John Sousoulas—owner of not one, not two, but three—restaurants along Route One in Connecticut (the third is in Mystic) sure does.

That very memory was the catalyst for the cataclysmic Mystic French Toast (recipe below), something that should totally be ordered for dinner (or when hungover). Or—yeah, whatever—breakfast.

The white bread is rolled in cornflakes prior to being deep-fried, after which is is smothered in strawberries, bananas, and whipped cream, not to mention stuffed with cream cheese. A meal perfect for lazy days to be sure.

The Parthenon—as it is known by townies—runs the gamut of diner fare, up to and including a killer chicken parm, plus adventurous dishes such as the Chicken Sensation, which finds said chicken keeping company with artichokes, roasted red peppers, and mushrooms in a white wine sauce. Begin a day here or end one here; you won't be sorry either way.

SOUNDTRACK SUGGESTION

"Born to be Wild" by Steppenwolf

I'm really not of the mindset where playing "Born to be Wild" needs explanation, ya know? It fits most any occasion, but when a group of cronies who are famished take it to the diner, this is the song that comes to mind.

Mystic Stuffed French Toast
SERVES AS MANY AS YOU NEED

FOR THE BANANA PUREE

bananas

cream cheese

vanilla extract

sugar

FOR THE RUM SAUCE

cups heavy cream

sugar

rum

FOR THE FRENCH TOAST

4 eggs

2 slices bread

fresh strawberries, sliced and divided

cornflakes, crushed

olive oil

bananas, sliced

raspberry syrup

For the banana puree

Blend bananas, cream cheese, vanilla extract, and sugar until smooth. Set aside.

For the rum sauce

In a medium saucepan over low heat, mix heavy cream, sugar, and rum and cook slowly until thick.

For the French toast

In a small bowl, crack and whip eggs.

Spread banana puree on 2 slices of bread, add half of the sliced fresh strawberries. Put bread together like a sandwich and dip in whipped eggs. Roll sandwich in crushed corn flakes and freeze for 1 hour.

When ready, heat oil in a deep pan. Deep fry the corn flake-encrusted sandwich until golden crisp. Cut the sandwich in half and top with rum sauce. Garnish with sliced bananas and more strawberries. Drizzle with raspberry syrup and enjoy.

CHEF'S TIP

You can make your own Stuffed French Toast at home, but, it's a lot easier to just come in, relax, and enjoy it at the Parthenon Diner!

SHORELINE PRIME MEATS & DELI

103 N. Main St., Branford CT
shorelineprime.com, (203) 208-1579

Lawyer by day . . . butcher at lunchtime? New Haven attorney Albert Carocci has always had a hand in the food biz in one way or another (he was once part-owner of the Elm City's venerable Fireside Grill), but Shoreline Prime is a total labor of love. Initially just that—a tried and true deli—it has since grown to a sizable grocery story, stocked to the rafters with specialty items largely produced by Connecticut farmers and entrepreneurs. There are a few tables in there now, prepared foods for those on the go, and the catering business is booming.

But let's talk sandwiches.

There is a Grilled Mac 'n Cheese to die for and, no, you're not misunderstanding. It's an epic grilled cheese, loaded with creamy mac 'n cheese. There is also the "Thimble Island," (a chicken cutlet with bacon and cheddar named after the nearby Branford destination) and a pulled pork sammie called the "Saucy Li'l Piggy." I don't know who that's named after.

Shoreline is best known for its award-winning titular steak and cheese sandwich. You can purchase the bourbon-soaked tips used in the sandwich if you'd prefer to grill 'em on your own.

Prime's offspring is Al's beloved "Silver Bullet," a food truck that careens up and down Route One, where he can often be see doling out the steak and cheese outside sporting events, Gatsby-like shindigs, and more.

SOUNDTRACK SUGGESTION

Warren Zevon's "Werewolves of London"

Shoreline is quaint yet proper enough to give off a London vibe, so Zevon's classic is perfect— especially as the only thing that can take out a werewolf is . . . a silver bullet, of course.

FRIENDS & COMPANY

11 Boston Post Rd., Madison, CT
friendsandcompanyrestaurant.com, (203) 245-0462

Sure, there is more than one restaurant called Friends & Co., so it is up to each individual owner to set his or her establishment apart from the others. Consider that mission accomplished in Madison. Established in 1980 on the Connecticut shoreline between the Madison and Guilford border, Madison's Friends & Company offers a great view of the East River and while it predates the venerable NBC sitcom, it could have easily displaced that show's Central Perk. This is where you go with your crew to celebrate a birthday or someone getting a raise or even that divorced finally coming through. It's that place.

SOUNDTRACK SUGGESTION

"I'll Be There For You" by The Rembrandts

(yeah, the Friends theme song)

Punchy, poppy, and perfect, this song stands on its own, even while it will always conjure up images of Ross, Rachel, and the gang. Friends & Co. is totally the type of place they would have inhabited in their wild and woolly twenties. If their coffee shop closed for renovations, that is.

One menu staple is the vegetable fritters, a bounty of deep-fried goodness served with a sesame soy dipping sauce that enables the fritters to explode in your mouth. Back in the '90s I was a sucker for any of the burgers on the menu because they came nestled on an English muffin. The muffin is gone nowadays (though you can still get it served that way should you request it), but the burgers remain.

I recommend the Italian Burger, equipped with arugula, roasted tomatoes, fresh mozz, and pesto.

The seafood is also top-notch, specifically any shrimp dish, with the seasonal clam fritters being a meal unto itself, as chock full of clam as it is corn. Boom!

Key Lime Pie
MAKES 2 PIES

12 ounces butter, melted

13 ounces graham cracker crumbs

¾ teaspoon ginger

¼ teaspoon nutmeg

8 eggs, separated

2 (14-ounce) cans sweetened condensed milk

1 cup key lime juice

⅛ teaspoon cream of tartar

¾ cup sugar

1 lime, sliced thin

Heat oven to 300°F.

In a medium bowl, mix butter, cracker crumbs, ginger, and nutmeg. Press the mixture into two pie pans and bake for 4 minutes.

In another bowl, beat egg yolks. Add milk and beat until light in color. Add lime juice and pour into cooked pie crusts.

Return to oven and bake until set, approximately 20 minutes.

Meanwhile, in another bowl beat ¾ cup egg whites to soft peaks. Add cream of tartar, and slowly add sugar until stiff. This will make enough merengue for 2 pies.

Spread ½ of the merengue on each pie, sealing it to crust. Bake for 3 to 4 more minutes.

Chill before serving

Garnish with lime

THE LOG CABIN

232 East Main St., Clinton, CT
thelogcabinct.com, (860) 669-6253

Look, my name is Vinnie. I'm Italian. If my father were still alive he'd be livid that I ordered the veal parm at a restaurant with such a decidedly American name. But I love me some veal parm, and The Log Cabin is so homey, you feel so comfortable, you just trust them. Trust me.

The Baked Mariner's Platter might also strike some as an ambitious order, especially as the place is literally surrounded by seafood joints (we are talking about the Connecticut shoreline after all), but do it. The plate is actually heavy, and the assortment of seafood assembled is to die for. It's just, with a name like The Log Cabin, you'd think this restaurant would be all about meatloaf and pot roast, right? Those dishes are there, but everything else is too. You eat hearty, and it's all prepared with heart.

SOUNDTRACK SUGGESTION

"Pink Houses" by John Mellencamp

Perhaps better cued up upon exit, a rousing chorus of "Ain't that America" after a feast at The Log Cabin pretty much sums up the entire experience.

The Chicken Jennifer is both chicken and shrimp at once, tender and delicious, finished with a demi-glaze that can get even the most road-weary co-pilot to forgive you your trespasses.

Chicken Jennifer
SERVES 3

3 4-ounce chicken breast (pounded out)

8 21/25 shrimp, peeled, deveined, and tail off

4 Medium asparagus blanched and cut

½ cup mushrooms, sliced and sautéed

5 slices mozzarella cheese

½ stick butter

2 oz white wine

1 fresh lemon

1 shallot, diced

Flour, to coat chicken

In medium sauté pan add olive oil, brown the chicken and add shallots, mushrooms, shrimp, asparagus, and white wine, plus juice from lemon and half stick of butter. Place in oven, and heat to 450°F for 15 minutes or until shrimp are cooked and cheese is golden brown.

ANTHONY J'S BISTRO

6 Holmes St., Mystic, CT
anthonyjsbistro.com, (860) 536-0448

About as far from Route One as we are going to stray—and well worth the extra minutes—Anthony J's is one of those joints that serves your meal sizzling atop sizable stones. In fact, this is how a lot of people refer to the place: "Have you ever been to Anthony J's?" "You mean the place with the stones?"

This portion of their menu is the "Hot Rocks" section. From chicken to steak to seafood scampi (Atlantic sea scallops, jumbo shrimp, calamari, Canadian lobster meat, and fresh fish), every dish comes prepared on imported stones from Italy that are heated to 650°F. You finish preparing the meal yourself at your table, where the soothing sound of food sizzling collides with that of the seagulls and nearby schooners honking their horns.

Elsewhere on the menu you can find staples but with a spin, such as the veal meatballs finished with a San Marzano tomato sauce, or garlic bread with so much cheese melted on top it's a meal in and of itself.

SOUNDTRACK SUGGESTION

"Dirty Deeds Done Dirt Cheap" by AC/DC

Angus Young's inimitable guitar style—especially what he does on this hit for the Australian rockers—is hot rocking in advance of the hot rocks.

Here's the deal: Anthony J's is the place that gets the lady on your arm talking to you again after she stopped an hour or so earlier on the road.

Penne and Scallop alla Vodka with Sun-dried Tomatoes

MAKES 2 SERVINGS

FOR THE VODKA SUN-DRIED TOMATO BASE

½ cup olive oil

1 tablespoon chopped garlic

1 tablespoon chopped shallots

5 basil leaves

1 cup canned plum tomatoes

1 cup sun-dried tomatoes

1 cup vodka

FOR THE PENNE AND SCALLOPS

1 tablespoon butter

½ cup sliced mushrooms

¾ cup chicken stock

1 cup heavy cream

⅛ cup peas

1 heaping tablespoon vodka base

6 Scallops

½ pounds penne, cooked half-way

Salt and pepper to taste

For the vodka sun-dried tomato base

Add oil, garlic, and shallots to pot. Sauté until garlic floats. Add basil, plum tomatoes, and sun-dried tomatoes. Cook over low heat for 5 minutes, stirring occasionally. Add vodka and bring to a boil. Remove from heat and purée.

For the penne and scallops

In a 10- or 12-inch sauté pan, combine all the ingredients except the penne. Stir and bring mixture to a boil.

Reduce heat and add pasta. Stir the pasta in until it is just coated by the sauce, do not allow it to become soupy. If you cook it too much, you can add a little stock to bring it back. Check for flavor, adding salt and pepper as needed.

Dinner and a Movie

Again, whether you are seventeen or seventy-seven, there is no better date night than the tried and true dinner and a movie night. Ticket stubs are oft-kept, the meals ordered memorized, outfits worn seared into memory. So, here we offer you yet another arthouse cinema awash in history and charm. Concessions here are curious, as in candy you'd find nowhere else versus standard-issue Milk Dud fare, and fliers for xylophone concertos are plastered on the wall of the lobby.

Mystic Luxury Cinemas
Olde Mistick Village
mysticluxurycinemas.com
(860) 536-4227

It seems as if this gem of a movie house has revamped itself to some degree over the past few years, but it simply cannot shed its history-laden charm. How could it when it is nestled inside a veritable village that screams yesteryear? Still, the word luxury has been plugged into its name, even while all the arthouse films that film-loving New Englanders hope won't remain "in limited release," off in Brooklyn, are still in tow. The surrounding shops, duck ponds, and eateries have been designed to represent a New England village of about 1720. In short, even if this little four-screening house decided to roll with the block-busters, the charm still could not be lost on its customers.

MINDY K DELI & CATERING

1610 Boston Post Rd., Old Saybrook, CT
mindyks.com, (860) 399-6427

Old Saybrook—aka the late Katherine Hepburn's 'hood—is perhaps only fathomable as a once-great Hollywood star's seafaring getaway by actually seeing it. I could bust out the thesaurus for all of the right words in an effort to convey to you just how idyllic, quaint, and artistic this Connecticut suburb is, but it would pale in comparison to you parking your car on the Boston Post Road and ambling around one summer afternoon.

SOUNDTRACK SUGGESTION

"Hungry Like The Wolf" by Duran Duran

Visions of a very young Simon LeBon traipsing through the forest in search of something to eat—and love—will dance through your head here, especially if you study up on the story gracing the walls. But the menu will take you to this '80s classic too.

Surely Mindy of Mindy K's gets my drift. Last name Khamvongsa, she and her sister-in-law Tippy are from Laos. In fact, as recent as 2012 they raised $10,000 right in Old Saybrook to build a school for poor children in their homeland. When they go home to Laos they bring gifts for blind children, and photos of these moments grace the walls of their charming, humble deli.

It's standard deli fare on the menu—with a grilled pastrami reuben to die for, but if you're in the mood for Thai, you are truly in luck; it is what really sets Mindy's apart. The Pho Noodle Soup will cure whatever ails ya, including a broken heart.

Tom Yum Goong Soup (Coconut Sour Spicy Shrimp Soup)

SERVES 2

2 cups chicken stock.

5 slices galangal root

5 Kaffir lime leaves

5 (1-inch) slices lemon grass

2 teaspoons fish sauce

½ small onion, cut into slices

1 tomato, cut into slices

5 mushrooms, cut in half

Salt to taste

10 to 12 shrimp

1 to 2 red chilli peppers, sliced down the middle

½ 13.5 ounce can coconut milk

Lime juice (to taste)

Cilantro to taste

In a sauce pot, bring chicken stock to boil over medium heat. Add galanga, lime leaves, lemongrass, and fish sauce. Let the mixture come to a boil and cook until you smell the lemongrass.

Add in onions, tomatoes, and mushrooms. Bring mixture to a boil again and begin adding salt for taste.

Add shrimp and chilies.

Stir in coconut milk and add lime juice and cilantro to taste. Additional fish sauce can also be added if desired. Continue to boil and adjust taste until you are satisfied with the end product.

Pahd Thai Gai (Chicken Pahd Thai)
SERVES 6

1 (12-ounce) package rice noodles (should be similar to linguini)

4 eggs

Salt and cracked pepper to taste

1 cup vegetable oil, divided

1 pound boneless skinless chicken breast halves, cut into bite-size pieces

⅛ tablespoon crushed red pepper

¼ cup crushed peanuts, plus more for garnish

3 tablespoons white sugar

2 tablespoons fish sauce

1 cup tamarind sauce

4 Asian onions or shallots, chopped fine

2 cups bean sprouts

1 lime, cut into wedges

Soak rice noodles in cold water for 30 to 50 minutes until soft. Drain, and set aside.

Crack eggs in a small bowl. Scramble with salt and cracked pepper to taste. Remove from heat and set aside.

Heat ½ cup oil in a pan over medium-high heat. Add chicken and sauté until browned. Add salt and cracked pepper for taste. Remove from pan and set aside.

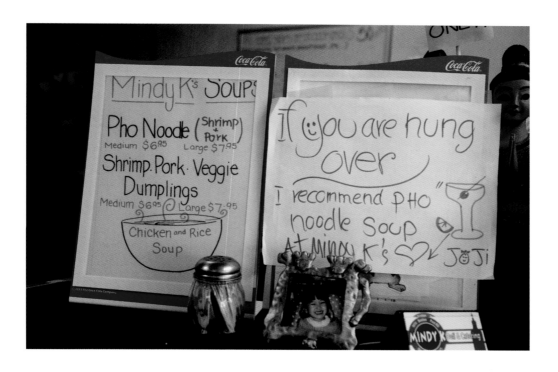

Heat the other ½ cup oil in pan, still over medium-high heat. Add noodles, cooked chicken, red pepper, ¼ cup crushed peanuts, sugar, fish sauce, and tamarind sauce.

As the noodles begin to sizzle and take color, add scrambled eggs, Asian onions (or shallots), and bean sprouts. Continue to stir fry until noodles are tender. Remove from heat, garnish with a slice of lime, and extra peanuts.

CHEF'S TIP

Feel free to add any cooked vegetables to your noodles to personalize this dish. Add them when you add the scrambled eggs.

JOHNNY SALAMI'S WORLD CLASS

205 Food Terminal Plaza, New Haven, CT
johnnysalamis.com, (203) 777-7906

It sounds like a Joe Piscopo movie, doesn't it? The tongue is planted firmly in cheek at this meat district–esque deli that sits squarely between a high-end Italian restaurant and the legendary Long Wharf Theatre, where luminaries such as Pacino and Streep have been known to take in a show. Oh, and it's a food terminal, too. Salami's is about as Route One as you can get in New Haven, where the "Q Bridge" abuts our strip o' road if only for a mile or two.

Every imaginable variation on the sandwich is the plunking of a ten-spot away, with my particular fave being the Ghetto Burger, as opposed to the White Trash Burger. (I told you the tongue was planted firmly in cheek here.) The former is maxed out on chili and cheese and demands three napkins, minimum. The Cajun chicken also comes highly recommended.

SOUNDTRACK SUGGESTION

"Heaven" by Warrant or Bryan Adams

With the hum of both I-95 and I-91 in the distance, some '80s light rock is the right accompaniment—believe you me—when you step through the doors of the rockin' Salami, and with so many portions of the menu talking Heaven, I'm leaving the choice between Adams or Rock of Ages Broadway show soundtrack up to you.

There are World Class Steak Sandwiches, too. No really, that's what they're known as on the menu—World Class. The chipotle steak on a roll will do ya until well past dinner.

The fun on the menu doesn't end there, though. There are more burgers in the "Burger Heaven" portion of the menu, and other choices in the "Panini Heaven" portion. Hyperbole? You decide.

LUIGI'S RESTAURANT

1295 Boston Post Rd., Old Saybrook, CT
luigis-restaurant.com, (860) 388-9190

Since 1956, baby. A visit to Luigi's will make you talk like Sammy Davis, Jr. It just happens. Boastful that their cuisine has roots "deep in the Old Country" and describing the dining experience as "red sauce eatin'," Luigi's offers unapologetic, unbutton-the-top-button feasting in a cozy, homey atmosphere, with all the Scorsese trimmings as far as the decor.

SOUNDTRACK SUGGESTION

"Layla" by Derek & The Dominoes

While dining at Luigi's, you can easily feel as if you are sitting in a room where some wise guys made decisions once upon a time. Luigi's is a spot where Scorsese could easily place Pacino or DeNiro (or Pesce!) and they'd feel right at home. To that end, Clapton's familiar licks here lend themselves to a "method dinner."

Celebrating sixty years in business in the summer of 2016, these guys know what they're doing. In fact, the guy who started it all "went to the grave," according to grandson Len, "saying he was the one who came up with the word grinder (for sandwich)."

But the pizza! Luigi's is keeping up with the Joneses pizza-wise, with some ambitious 2.0 toppings. That said, you'd be crazy to go here and not get straight-up traditional when it comes to ordering. This place will make you miss your Nana.

Oscar-winner Art Carney was a regular, and when my family and I began bandying about our favorite Honeymooners lines, my ten-year-old leaned over and whispered in my ear, "I can't believe Ed Norton used to come here. Dad . . . The Hulk!"

The 6-Cheese Italian Purses are my go-to order, the handmade pasta is formed into purse shapes, filled with six cheeses, and then tossed in a rich, creamy Alfredo sauce. The devout devour the seafood here, especially the jumbo shrimp scampi. Either way, that top button's coming undone.

Luigi's Chicken Leonardo

MAKES ONE LARGE PORTION, LARGE ENOUGH FOR TWO TO SHARE

1 (8- to 10-ounce) chicken breast

Flour for dredging

3 rounded tablespoons garlic butter

¼ cup sliced hot cherry peppers (marinated in a vinegar brine)

⅓ cup roughly diced roasted sweet red peppers, skins removed

½ cup white wine

½ to ¾ cup chicken stock (canned chicken broth may be substituted)

¼ pound angel hair pasta

1 tablespoon freshly squeezed lemon juice

Pinch kosher salt

Pinch fresh ground black pepper

1 lemon

½ cup blanched fresh baby spinach

Freshly grated Parmigiano Reggiano or other similar cheese

chopped parsley for garnish

Slice chicken breast horizontally into 2 large, flat pieces. Pound lightly.

Lightly salt a pot of water and begin heating to a boil for the pasta.

Heat garlic butter in a sauté or fry pan over medium-high heat.

Dredge chicken pieces through flour and carefully drop into hot garlic butter. Cook for approximately 1 minute.

Flip chicken over and add hot and sweet peppers. After approximately 1more minute, add wine. Let simmer for 1 minute, then add ½ cup chicken stock. Simmer until chicken is cooked through, adding chicken stock as needed to maintain some juice in the pan.

Drop pasta into now boiling water. Cook according to directions until al dente.

Add lemon juice, salt, and pepper to simmering chicken, then sprinkle spinach in pan and continue to simmer only until heated thoroughly.

To make the lemon twists, trim both ends of the lemon. Using a vegetable peeler or a sharp knife, work your way around and down the lemon to remove an unbroken spiral, trying to get as little of the pith as possible. Set aside for garnish.

Drain pasta, place on plate, and top pasta with chicken and contents of pan.

Sprinkle with grated cheese to taste and garnish with parsley and lemon twists.

Buon Appetito!!

Luigi's Famous Party-size Antipasto Salad Recipe
SERVES 15 TO 20

2 heads romaine lettuce, washed, chopped

4 ounces each cooked salami, Genoa salami, ham, capicola, sliced thin, provolone

1 small jar each of marinated mushrooms, marinated eggplant, pepperoncini, marinated artichoke hearts, roasted red peppers

2 tomatoes, cored and sliced in wedges

1 cucumber, sliced

15 to 20 slices pepperoni

15 to 20 pitted olives, (black, kalamata, or both)

2 (4-ounce) balls fresh mozzarella, sliced

4 links Italian sausage, cooked, sliced

1 small tin Anchovies in olive oil (optional)

oregano

granulated garlic

black pepper

2 cups Italian dressing

4 tablespoons extra-virgin olive oil

Mound chopped romaine lettuce on a 16-inch round platter (preferably a platter with a rim to prevent dressing from spilling).

Form each meat and provolone slice into a cone by cutting a slit halfway through and rolling into a cone, or by simply folding twice and opening into a cone. Insert cones randomly into lettuce.

Drain marinated vegetables and pepperoncini of their jar juices, set red peppers aside, and sprinkle rest over entire top of salad, covering mostly lettuce and keeping rolled meats/provolone exposed.

Evenly spread the tomato wedges, cucumber slices, and pepperoni over the top.

Place olives and fresh mozzarella on top, placing some of them in the meat and provolone cones.

Spread sliced Italian sausage on top of the salad.

Dice roasted red peppers and sprinkle evenly over the top.

If desired, lay anchovy filets over the top, or serve separately on the side.

Sprinkle generously with oregano, black pepper, and granulated garlic.

Drizzle antipasto generously with Italian dressing.

Serve with crusty Italian bread and enjoy!

———

CHEF'S TIP

The ingredients and amounts included here are Luigi's recommendations, but any similar Italian deli meats, cheeses, and vegetables may be used, in any amount you prefer. They say, "Be flexible, use what you like and have fun with it!"

ENGINE ROOM

14 Holmes St., Mystic, CT
engineroomct.com, (860) 415-8117

Some people like to climb off a boat, or out of a sports car, and grab a hot dog on the side of the road, while others prefer fixing themselves up to pate away the road. The Engine Room clearly tar-gets the latter, with its Manhattan-esque din-ing area and to-die-for dining experience.

Whether it's buttermilk-fried squid and green goddess aioli or a roasted Brussel sprout and kale salad with country ham, it's somehow Nashville-by-way-of-Manhattan tucked away in Mystic, Connecticut. Three state lines in one shot.

The Engine Room buys local, and why wouldn't they when their options include Mys-tic Cheese Company, Narragansett Bay Lob-ster, and Maple Lane Farms, to name but a few? They also work closely with

Beriah Lewis Farm of North Stonington, right up the road, and purchase a whole cow every week.

Having restored a marine engine building (hence the name) for their restaurant, the players here are not interested in doing anything halfway. Nor should you be when it comes to feeding yourself!

MONKEY FARM

571 Boston Post Rd., Old Saybrook, CT
themonkeyfarmcafe.com, (860) 388-4866

The Monkey Farm looks like a big old house onto which somebody just slapped a neon sign a few scant feet above an "Open for Business" one. What's more—it feels that way, too.

A guy named Harry Corning bought the former inn in 1968, and today his three kids run the place. Again, you somehow "feel" all this while having a bite and throwing a few back in here. A favorite of the locals, the food—yes, I'm going there—hangs off the plate. In the case of the roast beef sandwich, it is stacked high enough to take the ceiling fan into consideration.

The guy behind the bar'll tell ya to go with the fresh tuna salad; that's your decision. I'm partial to the French onion soup, with the blueberry bread pudding for dessert. Stick around and enjoy the band—you're not going to get right up anyway.

SOUNDTRACK SUGGESTION

"Wild World"
by Cat Stevens

The Farm's got a throwback vibe happening with oodles of history permeating the walls. And the regulars know how to play right into all of that, rendering Stevens' penultimate hit—and chant—a righteous rejoinder.

LITTLE BARN

1050 Post Rd. East, Westport, CT
littlebarnct.com, (203) 557-8501

Little Barn hasn't been around forever, but that doesn't mean it won't be. Established in 2013, it is an unassuming spot in affluent Westport, Connecticut, with a thing for tacos.

The mahi tacos will put the island in a city girl, and the beef tacos are just straight-up cowboy good, char-grilled filet tips loading the soft shell. That's a lot of imagery, huh? At a reasonable $3.95, it's a lot of tastes good, too.

Little Barn definitely doesn't skimp when it comes to the carnivore, with one burger on their sizable burger menu topped with bacon and egg, and the steak frittes a good size and cooked to perfection.

As quaint as it is competitive, Little Barn is an eye-popping entity—either restored from a former glory, or made to appear all yesteryear. Does it matter? Either way, it's welcoming, and all about you coming back now, ya hear?

SOUNDTRACK SUGGESTION

"Hold My Hand" by Hootie & The Blowfish

This place is ripe for a hootenanny, and who better than Hootie, especially front man Darius Rucker himself, to be in charge, turning the barn into a concert hall one verse in?

THE PLACE

901 Boston Post Rd., Guilford, CT
theplaceguilford.com, (203) 453-9276

The beauty's in the concept at The Place. The whole dang thing is an outdoor event—seashells crunching beneath flip-flopped feet, food cooked over an 18-foot fire pit, customers eating on tree stumps. Way to save on the overhead, right? Well it works.

Get ready for a crowd come Friday night in the summer, because The Place is a force to be reckoned with on an increasingly competitive stretch of Route One. Just one glance out your car window at their boisterous crowds will compel you to pull said car over, wait be damned. If it looks like it's a party, you can be sure as hell that it feels like one, too.

It's a glorified picnic, and the roasted corn actually rivals the smoked lobster, it's that good. Buttery, tender, and juicy, it's as much a staple of summer as any seafood dish, and The Place is savvy enough to know this. The plain roast clams are a must as well. You're also welcome to bring your own sides and—you guessed it—a bottle, too.

SOUNDTRACK SUGGESTION

"Cheers to That"
by Rihanna

The Place is a party. It's a backyard BBQ in no one's backyard, where everyone's in it together, and to that end one massive toast goes down well, sitting beside dozens of complete strangers. Set it to music I say.

TRE SCALINI RESTAURANT

100 Wooster St., New Haven, CT
trescalinirestaurant.com, (203) 777-3373

Tre Scalini can be a great place for celebrity watching, thanks to its proximity to the world-renowned Long Wharf Theatre, which is less than a minute away and known to boast productions that either star or attract A-list celebrities. I once passed Pacino on my way out of Tre Scalini.

SOUNDTRACK SUGGESTION

"Oh What A Night" by The Four Seasons

Anything with Frankie Valli out front captures the spirit of Tre, with its quaint corners and chandeliers. But, interestingly, this hit, where the other guys in the band are front and center, captures the spirit and the energy.

Known for the kind of traditional Italian fare a Corleone would love, and with an ambiance on par with a place on Mulberry, my go-to is the vitello (veal) milanese, a cutlet pounded out so beautifully that you gotta "tweet it before you eat it." The osso bucco is another gem, the pork shank is roasted so slowly as to make it workable with a spoon, entangled with a creamy parmesan risotto.

Tre Scalini's soups are worth mentioning, too, with their pasta fagioli the standout. Or the escarole and beans. In fact, order up some cutlets nestled on a plate of escarole and beans. They won't mind.

Paccheri Portofino

SERVES 8

2 tablespoons olive oil

½ pound of pancetta, cut into ¼- by ¼-inch cubes

1 medium-size onion, finely chopped

Hot pepper flakes to taste

Salt and pepper to taste

1 28-ounce can of San Marzano whole-peel plum tomatoes in purée

2¼ pounds Paccheri pasta or large rigatoni

⅛ pound pecorino Romano cheese

In a large sauté pan on medium heat, add olive oil and cubed pancetta and cook until pancetta is almost completely rendered and has a rich caramel color, 6 to 8 minutes.

Add chopped onion and sauté for another 3 to 5 minutes until onion is translucent.

Add red pepper flakes to taste and a pinch of salt and pepper (more will be added just before finishing).

Crush San Marzano tomatoes with hands until all are broken up. Add to onion mixture and cook for 20 minutes at a slow/medium simmer.

Meanwhile, cook pasta according to package instructions.

Lower heat to low and add fresh cooked pasta. Sprinkle with Romano cheese to taste, keeping in mind that the pancetta and cheese are salty. Add more salt and pepper if necessary.

Toss and serve.

PAUL'S PASTA SHOP

223 Thames St., Groton, CT
paulspastashop.com, (860) 445-5276

Paul's Pasta Shop was opened in 1988, the same year Bon Jovi released their *New Jersey* record (apropos of nothing really), by owners Paul and Dorothy Fidrych. The couple had been married for four years or so, wanted to raise their family in the area, and wanted moreover to simply make great pasta while gazing at the Thames River. Mission accomplished.

Paul's will sell you the pasta for you to boil yourself and take credit for at home, or even cater a huge family event, but the beauty's in the booth, man, where a great selection of beer or wine, and a literal vacation from the kitchen awaits.

My particular fave? The eggplant parm served over the pasta of your choice. For me it's angel hair or linguini, as tender and tasty as can be. The spaghetti with sausage and peppers is a close second, though.

SOUNDTRACK SUGGESTION

"I'll Be There For You" by Bon Jovi

Since the restaurant is technically a love letter from husband/owner to wife/owner, the music you play while heading over should be a love letter as well. And what better than this bombastic Bon Jovi ballad—released the very year the place opened?

Pauls's shared their delicious creamy garlic dressing—perfect on a salad to accompany your pasta.

Creamy Garlic Dressing

MAKES 2½ PINTS

1 quart extra heavy mayonnaise

6 ounces water

3 ounces chopped garlic

2 ounces lemon juice

1 ounce chopped parsley

⅛ teaspoon salt

⅟₁₆ teaspoon black pepper

Combine all ingredients in a bowl and stir. Season to taste.

JACK RABBITS
(BURGERS, WINGS & THINGS)

254 Main St., Old Saybrook, CT
jackrabbitsct.com, (860) 510-0048

Because sometimes a pit stop has less to do with eating and more to do with human interaction. This is one of those places where everybody knows your name. The energy is contagious here, and while you may drag yourself into Jack Rabbits, you'll sure as hell find yourself hopping out.

They lobster-fy pretty much everything here, from the hot dogs to the sliders to the BLTs, but it's the wings that put this place on the map. Dirty, Polynesian Orange BBQ, Honey Mustard . . . any way you want them that's the way you can get them. They also do hot dogs a hundred different ways and take the "gourmet" in "gourmet burger" deadly serious. Try the Wellington with its seared foie gras, mushrooms, and grilled onions or the Pepe Le Pew, resplendent with grilled apples and brie.

On second thought, maybe you won't exactly "hop" outta there.

SOUNDTRACK SUGGESTION

"Any Way You Want It" by Journey

Has this Journey mega-hit ever been used in a restaurant commercial before? I mean, it should. Especially for places that stand by such ideology: ANY way you want it, THAT'S the way you need it. Jack Rabbits certainly does.

Here are three variations on wings Jack Rabbits-style:

Siracha Lime Wings

½ cup water

½ cup fresh-squeezed lime juice

½ cup siracha

chopped cilantro for garnish

Polynesian

1 cup orange marmalade

1 cup ketchup

½ cup red wine

Jack Rabbits Magic Sauce

1 cup mayonnaise

1 ounce A-1 steak sauce

2 ounces ketchup

MOE'S BURGER JOINT

997 Main St., Bridgeport, CT
moesburgerjoint.com, (203) 333-9500

Moe hit Bridgeport in 2010, and the town hasn't been the same since. Harkening back to a kinder, gentler era, the decor is all Arnold's from *Happy Days,* or even the place where Danny and Sandy shared a burger in *Grease* before Kenickie got a shake dumped all over him. Why would you ever waste a perfectly good milkshake?

SOUNDTRACK SUGGESTION

"You're The One That I Want" from the *Grease* soundtrack

Since the place can double for the spot where Frankie Avalon's Teen Angel showed up to serenade Frenchie, this number-one hit from 1978 is the obvious choice.

The burgers at Moe's are top-notch, from the Bridgeport with it's grilled onions, bacon, lettuce, and horseradish mayo to the Juicy Lucy and it's double patty stuffed with bacon and cheese, fried pickles, and Moe's own sauce. And they sure know how to smother here, the hot dogs get ample chili, the burgers get mac and cheese (you read that right), and the fries are practically buried in bacon and cheese, but it's all good.

In short, Moe's knows.

Moe's Slaw Recipe

SERVES 8

1 head cabbage, shredded

1 whole carrot, shredded

2 ounces red wine vinegar

2 teaspoons sugar

1 teaspoon oregano

1 teaspoon garlic power

1½ cups mayonnaise

Place shredded cabbage and carrot in a medium bowl, and mix in all other ingredients until well mixed. Chill and serve as a side or as a topping on your favorite sandwich.

This cole slaw is an amazing recipe for topping any chili dog, chili burger, jerk chicken sandwiche, or pulled pork. Have fun and explore with food.

FLIPSIDE BURGERS & BAR

Brick Walk, 1125 Post Rd., Fairfield, CT
flipsiderestaurant.com, (203) 292-8233

Flipside Burgers & Bar is a literal hideaway in an alley just off the Post Road in upscale Fairfield County, but people from all around seem to be finding a way to get to this place.

It's a mix-and-matchers paradise, wherein you can customize your own burger. Want 6 ounces of sirloin with pepper jack and potato chips? Go for it. Maybe a turkey burger with baby swiss and guac is more your style? Just say so. You can do the same with the hotdogs.

Or you can go rogue and tackle myriad mac and cheese incarnations. The Land & Sea is loaded with chicken, shrimp, and mushrooms, while the Tex-Mex is weighed down with fresh pico, barbecue chicken, and potato chips.

The sandwich board is equally ambitious, with the Italian Philly being the real standout. It's a Philly cheesesteak by way of Sicily.

Flipside is atmosphere with a side order of good eatin'.

SOUNDTRACK SUGGESTION

"Go Your Own Way" by Fleetwood Mac

Heed Lindsay Buckingham's advice and go your own way, both to this hideaway, and once you're ready to place your order. Flipside celebrates individuality, as does this Top 10 hit from the Mac's *Rumours* record.

Caribbean Burger with Roasted Red Pepper Slaw

No exact measurements here—but you can try to get the flavors at home.
MAKES 1 BURGER

MANGO SAUCE

mango puree

onion

jalapenos

crushed red peppers

ROASTED RED PEPPER SLAW

shredded cabbage

roasted red peppers

pickled onion

vinegar

mango sauce

FOR THE BURGER

1 (6-ounce) sirloin burger

slice pepper jack cheese

1 hard roll

For the mango sauce

Blend all ingredients together and set aside.

For the roasted red pepper slaw

In a medium bowl, mix cabbage, peppers, and onions. Stir in vinegar and mango sauce until well combined. Chill until ready to serve.

For the burger

Cook the burger to your liking. Top with pepper jack cheese and allow to melt. Place cheeseburger on hard roll and serve with the red pepper slaw.

LYNN'S DELI

318 E. Main St., Branford, CT
lynnscaters.com, (203) 488-3232

Lynn's is your tried-and-true family business, a Branford institution. The DeMusis family has been serving the community for four decades, with the Lynn in question proudly handing over the restaurant's prize-winning mac and cheese recipe. "Written in my father's own handwriting," she gushes. She's right, the late patriarch, "Babe" DeMusis wrote, "Don't burn the cheese! Have a nice day" right there at the bottom.

Lynn's mac and cheese is creamy and satisfying, comfort food at its finest. Their chicken cutlets just might be second to none, the perfect amount of crispy, and sliced just so. In the summer you might get lucky and find squash flowers on the menu, a deep-fried goodness that can transport the rowdiest of roadsters to the most Zen of places.

Right on Route One, Lynn's is also crazy convenient to those traveling the interstate; it's maybe half a minute from both on- and off-ramps.

What's your excuse?

Mac and Cheese

4 (12-ounce) cans evaporated milk

2 quarts light cream

½ pound American cheese

½ pound Velveeta cheese

4 pounds elbow macaroni

Grated Parmesan cheese, to taste

Combine milk, cream, and cheese in a medium sauce pan and stir until cheese melts, set aside.

Cook pasta according to package instructions. Strain and add to sauce pan with cheese. Mix well and sprinkle with grated cheese. Remember, "Don't burn the cheese!"

G-ZEN

2 E. Main St., Branford, CT
g-zen.com, (203) 208-0443

G-Zen is our lone vegetarian/vegan restaurant in this book, and that's just fine. Quite frankly, this place is without peer. It's organic vegetarian sustainable cuisine kicks the catch phrases to the curb in favor of exceptionally good chow. In short, with most of the items on the menu you'd be hard-pressed to know the goal here is food that tastes good and is good for you.

SOUNDTRACK SUGGESTION

"California"
by Phantom Planet

or "California Dreamin'"
by The Mamas & The Papas

Located on a fork in the road in about as New England a spot as you can get, the atmosphere here is ALL California, from the decor to the food to the drink to two minutes of pleasantries with either of the owners.

The Zen Burger is just that: peace on a bun. The patty is erected from the tastiest of organic black beans and chipotle, and topped with your standard burger trimmings (lettuce, tomato, etc.; you take it from there). The Rock the Casbah is a divine traditional Moroccan stew. Their artisan cheese plate is straight-up second to none, house-made from cashews, all helmed by the owners from their bountiful farm only miles away in Durham, Connecticut.

At G-Zen they're taking "farm to table" to the next level. They're even got wellness cocktails on the menu. Ami Beach and husband, renowned chef Mark Shadle, consider themselves to be "wellness warriors." You'd be hard-pressed to argue.

Chocolate Raspberry Hazelnut Cake

SERVES 8 TO 10

FOR THE CHOCOLATE FROSTING

1 quart vegan chocolate chips

3 boxes firm silken tofu

3 tablespoons vegetable oil

1 cup organic sugar

¾ cup cocoa powder

2 tablespoons vanilla extract

FOR THE CAKE

1 cup vegetable oil

2½ cups organic maple syrup

1 teaspoon lemon juice

2 tablespoons vanilla extract

1 cup apple juice

3 cups organic whole wheat pastry flour

1 cup cocoa powder

3 tablespoons baking powder

½ teaspoon salt

1 jar (10 oz) 100 percent organic raspberry preserves, no sugar added

1 cup toasted hazelnuts, crushed

For the frosting

Melt vegan chocolate chips in a double broiler, stirring as needed.

Process tofu in a food processor until smooth and creamy.

Add in vegetable oil, sugar, cocoa powder, and vanilla extract. Process until smooth.

Add melted chocolate chips and process until smooth.

Let set until firm and then frost cakes when cool.

For the cake

Preheat oven to 350°F. Oil and flour three 8-inch cake pans.

Mix vegetable oil, syrup, lemon juice, vanilla extract, and juice in a bowl.

Add in flour, cocoa powder, baking powder, and salt, being careful to not over mix.

Pour equal amounts of batter into cake pans.

Bake on top of sheet pan for 20 minutes. Turn sheet pan and bake for 5 to 10 minutes longer.

Center should be springy and cake should pull away from sides when done. Cool in pans, then turn onto cooling rack to finish cooling.

Cut layers in half with serrated knife. Alternate spreading raspberry preserves, then chocolate frosting and chopped hazelnuts between all layers.

Garnish with remaining hazelnuts.

Gluten-free Raw Cacao and Coconut Truffles
MAKES ABOUT 2 DOZEN

3 cups organic shredded coconut

1½ cup raw cacao powder

1 teaspoon Ceylon cinnamon powder, plus more for dusting

1 teaspoon Lacuma Powder (optional but gives a rich caramel flavoring to truffle)

Pinch of Celtic sea salt

⅓ cup raw, cold-pressed coconut oil

¾ cup agave nectar, maple syrup, or sweetener of choice.

Add 2¾ cups shredded coconut, cacao powder, cinnamon powder, Lacuma Powder (if using), and sea salt into a large bowl, and mix by hand. Reserve about ¼ cup of the coconut for finishing touches.

In a small saucepan, warm the coconut oil just enough to liquefy. Add to dry mix along with sweetener of choice, and mix well with a wooden spatula.

Using a teaspoon or melon baller, scoop out raw chocolate mixture into small balls. Slightly dampen hands and roll into round truffle-size pieces.

Roll each truffle in extra shredded coconut, then dust each with a little Ceylon cinnamon for a final touch.

Put into fridge or freezer for 20 minutes to harden.

Remove them 10 minutes before serving and add a sprig of fresh mint and a dusting of cacao powder to the plate for garnish.

STONY CREEK MARKET

178 Thimble Island Rd., Branford, CT
facebook.com/Stony-Creek-Market-120336981312406,
(203) 488-0145

This is probably the farthest I ask you to stray from Route One, and it's completely worth it, especially if you order the market's legendary meatball sandwich. You can say "grinder," and even "hero" when ordering, but requesting the recipe for this will result in an eyebrow arching like you've never seen before, particularly if you're asking owner Valerie Wilkins, whose grandma is responsible for the ab-fab balls.

They are more forthcoming when it comes to everything else on the considerable menu, from omelets to pizza. Pizza nights are a community event at Stony Creek, with locals rushing over, bottle in tow (it's BYOB), and grabbing their spot early for tomato pies that will soon be heaved from the ovens.

SOUNDTRACK SUGGESTION

"All the Small Things" by Blink 182

This is one of those top-secret spots that the regulars take very personally, and pride is taken in every meatball made, every egg sandwich or pizza compiled. It's the small things that make this place loom so large.

Located directly across the street from Long Island Sound, the market also provides diners one of the best views in the book, a bird's eye view of a marina and boatyard, and sea air that pretty much makes road trips not only do-able, but enjoyable.

116 CROWN

I beg of you, hop off Route One and take on the faux Manhattan maze that is downtown New Haven. Then again, no I don't. I don't beg you anyway. But you really should check out 116 Crown.

Ostensibly a watering hole not all that different than the dozen others in the surrounding area that Yaleys belly up to, a visit to 116 Crown proves that it is much more than that.

Offering up veritable feasts at a fraction of the cost, this place also boasts a wine list arguably in the top five of this entire book, especially size-wise. The cuts of meat are equally sizable, grilled to perfection, and taken to the next level, should your arteries decide they are up for the challenge. Or peel it back and pair it with beets, rainbow carrots, and a porcini aioli. It's your call.

There is a raw bar, hand-cut charcuterie and cheese, and a bacon-wrapped pumpkin unlike anything you've ever had before. If taste-buds could send thank-you cards . . .

SOUNDTRACK SUGGESTION

"Holiday" by Madonna

The Material Girl's early hit fits like a jigsaw here. Her early years in New York City pulsating in this track and 116's surroundings capture that very vibe, never mind the "holiday" sensation that takes place every time your sizable entree is placed before you.

Burratta Panzanella

SERVES 4

10 cherry tomatoes, split

1 loaf Olive Batard, torn into bite-size pieces

2 teaspoons olive oil (Fratelli Colletti brand preferred)

1 teaspoon white balsamic vinegar

Pinch of salt

Pinch of pepper

3 local basil leaves, shredded

1 (2-ounce) ball burratta

Fry the pieces of Olive batard in 1 teaspoon of olive oil until brown. Add tomatoes, 1 teaspoon of olive oil, vinegar, pinch of salt, pinch of pepper, and basil leaves. Toss together and top with burratta.

Truffle Fries
SERVES 2

½ cup canola oil

3 potatoes, freshly cut into shoestrings

½ teaspoon salt

½ teaspoon pepper

1 teaspoon truffle oil

2 ounces Parmigiano-Reggiano cheese

1 teaspoon fresh chives

Heat canola oil in a deep sauce pan. Gently place fries in the hot oil and deep fry until golden brown. Remove to a large bowl and add salt, pepper, and truffle oil and toss vigorously. Place in serving bowl and top with cheese and chives.

Cookout
SERVES 2

40 ounces bone-in, dry-aged ribeye steak (tomahawk)

Salt and pepper to taste

2 ounces butter, divided

1 ounce brown sugar

6 ounces prosecco

7 baby carrots, split and blanched

10 ounces baby spinach

1 tablespoon olive oil, to coat pan

5 ounces heavy cream, divided

2 ounces Parmesan

4 Idaho potatoes

1 ounce butter

3 cloves black garlic

3 white onions, caramelized

Preheat oven to 425°F.

Heavily salt and pepper the ribeye. Heat a sauté pan to high heat. Place tomahawk into the pan and sear both sides.

After searing, place the ribeye in oven until cooked to proper temperature, about 25 minutes.

In a separate pan, melt 1 ounce butter with brown sugar and prosecco, reduce until it is a glaze-like consistency. Add blanched baby carrots and mix until carrots are glazed.

Sauté spinach with olive oil, salt, and pepper. Add 3 ounces cream and reduce. Finish with grated parmesan cheese.

Boil potatoes until soft, drain, and set aside.

In another pan, warm remaining 2 ounces cream and 1 ounce butter, and black garlic, salt, and pepper to taste. Add to drained potatoes and whip until creamy.

Heat caramelized onions on high heat.

Plate the ribeye alongside the glazed carrots, spinach, and potatoes. Top all with onions and serve. Enjoy!

PIZZERIA MOLTO

(Brick Walk) 1215 Post Rd., Fairfield, CT
pizzeriamolto.com, (203) 292-8288

I love this place. Replete with classic movie stills projected onto the walls, a Mulberry Street vibe, and first date/date night ambiance, Molto is totally Pizza Place 2.0.

The wine selection is spectacular, from Spain to Argentina to Tuscany to California, and syncing the right pizza up to the right bottle is a whole lot of fun. Try pulling it off with the Beet Pie, which features arugula, beets, goat cheese, caramelized onions, and a balsamic reduction to create a taste like no other.

Molto is equally ambitious with their tapas and paninis. The salmon BLT panini shatters the panini glass ceiling.

This place is located in a great spot to boot—bustling Fairfield downtown, with high-end shopping, a brilliant ice cream parlor, and a stellar live music venue all within walking distance. Bravo, Molto.

Pappardelle Scarpariello

SERVES 2

1 tablespoon olive oil

1 clove garlic

2 ounces crumbled Italian sausage

3 ounces chicken breast, half-cooked and sliced

¼ Spanish onion, sliced thin

2 cherry peppers, sliced thin

½ cup white wine

Touch of marinara for color

½ cup veal stock

1 tablespoon butter

Salt, pepper, and Italian herbs to taste

½ pound pappardelle, cooked to package instructions

Heat olive oil in a sauté pan over high heat. Add the garlic and cook until brown and aromatic.

Add sausage, chicken, Spanish onion, peppers, wine, marinara, stock, butter, and spices.

Sauté in pan until well combined.

Add cooked pappardelle and toss until mixed. Serve and enjoy.

Pizza My Mind

New Haven, Connecticut, is pretty much known for its pizza. Long-standing establishments such as Pepe's and Sally's have made national headlines time and again for their simple, mouthwatering pies. But, the fact is, this all got contagious years ago, and as a result there are literally dozens of pizza places throughout Connecticut that are fantastic, with plenty of 'em right on Route One.

So, let's hear it for the pizza joints we've featured, but let's also acknowledge the ones briefly listed here.

Brazi's Italian Restaurant
201 Food Terminal Plaza
New Haven, CT
(203) 498-2488

It doesn't get more Route One in New Haven than Brazi's, where you can actually watch the pizzas get hoisted into the brick ovens and hauled back out through the glass of the front door, like the kids in *A Christmas Story*. WARNING: The aroma is irresistible, and the staff so inviting they take "making you feel like family" to another level. Brazi's is so many important things all at once: Great Italian food at a ridiculously reasonable price, convenient, a landmark (even if it hasn't officially earned that status, it sure as hell is one), and with pizza as good as any of the big boys. Plus, Eddie is the greatest waiter not only in New Haven . . . not only in Connecticut . . . not only up and down Route One . . . but I'd say EVER.

Colony Grill
1520 Post Rd., Fairfield, CT
(203) 259-1989

With three other locations, and a rock-solid reputation, the hot oil bar pie is clearly where the oveneers (is that even a word?) hang their hat. Word of advice: Water, and plenty of it. Early drafts of the book did not include "The Colony," and when word got out there was an uprising, not unlike the dough itself. You have been heard. Word.

Grand Apizza Madison
734 Boston Post Rd., Madison, CT
(203) 245-8438

A New Haven pioneer got smart and took his wares to the shoreline, setting up shop right on Route One in Madison, and Madisonites and everyone passing through are all the better for it.

GG's Wood-Fired Pizza
591 Boston Post Rd., Milford, CT
(203) 876-8000

Despite being a full-scale restaurant that reeks of chain, GG's is old school enough to make its first offer on the pizza menu a Tomato Pie, which tells the pizza connoisseur (i.e. snob) everything he or she needs to know. Order one with soppressata.

Cappetta's Italian Imports and Pizza
188 Boston Post Rd., West Haven, CT
(203) 937-7485

When an establishment has "Italian Imports" in it's name, that typically means you can buy a block of cheese or olive oil if you're so inclined—a clear indicator that the pie will be solid. Try the Pizza Mariella, if only because the arugula that surfs the pie is a thing of freakin' beauty.

Palmieri's
316 Boston Post Rd., Waterford, CT
(860) 437-3730

Okay, okay, so Palmieri's wears its New York love on its parmesan-encrusted sleeve. So be it. It's one of the cooler-looking spots you'll spy on Route One, and their "We are not a chain and we'll never be a franchise" battle cry reminds me of Mel Gibson's final scene in *Braveheart*. Get the artichoke heart.

Tolli's Apizza & Restaurant
410 Main St., East Haven, CT
(203) 469-9582

At one point in time Tolli's occupied the same rarified air as Pepe's and Sally's, their name spoken with the same reverence, and deservedly so. Perhaps it's the East Haven versus New Haven end of the equation that cost it its mention alongside the greats, but the pie is still on par with them. Here, go for the escarole and beans pizza; it is to die for.

Marco's Pizzeria & Restaurant
313 E. Main St., Branford, CT
(203) 483-1544

Marco's has its Wooster Street ties, and you can taste it in the pies. But the proof is in the pizza.

Pizza Works Pies & Suds
455 Boston Post Rd., Old Saybrook, CT
(860) 388-2218

Pizza Works does it directly on stone and without oil. Talk about food porn! There is only one size pie made available and they source locally. A white pie is my go-to here, and while they certainly get ambitious with these (Tex Mex and Thai Chicken), I stick with the Margarita.

Food is never having to say you're sorry.

RHODE ISLAND

AL FORNO

577 S. Water St., Providence, RI
alforno.com, (401) 273-9760

When you pull up to Al Forno, you're gonna be taken seriously, plain and simple. This is the kind of place where you either propose or break things off for good, where you're either hired, fired, or promoted. If it's bad news, the recipient will cry significantly less tears, sighing the words "They got class" as they peel away from the curb after splitting the croque mademoiselle (recipe below).

Al Forno doesn't scrimp on anything. The silverware and decor are pristine, the wine selection second to none, and the food? Fuhgettaboutit.

There is an entire section of the sprawling menu dedicated to wood-grilled fare, from the decadent bacon-wrapped veal tenderloin to the tantalizing George's Bank scallops to— for you non-carnivores—the wood-grilled and roasted veggie entree.

SOUNDTRACK SUGGESTION

"Since U Been Gone" by Kelly Clarkson

The thumping bass line to this first *American Idol's* biggest hit is probably the only thing that can elevate an already extraordinary day, but doubles as a celebration of things over, which really hits the proverbial spot.

And if the handmade pasta doesn't say "I love you," the fresh-to-order desserts sure do. Hand-churned orange granita? Native rhubarb cobbler? Love is . . .

Croque Mademoiselle with Crème Anglaise

SERVES 6

FOR THE CRÈME ANGLAISE

1 quart heavy cream

1 quart milk

1 cups vanilla sugar

1 vanilla bean

1¼ cups pasteurized egg yolk

FOR THE CUSTARD

2½ cups heavy cream

3 eggs

¼ cup granulated sugar

2 teaspoons vanilla extract

FOR THE BRIOCHE

½ teaspoon of almond extract

amount confectioners' sugar, plus
more for dusting

½ cup mascarpone cheese

6 slices brioche

½ cup semisweet chocolate (chopped
finely)

For the crème anglaise

Pour the cream and milk into a heavy-bottomed saucepan. Sprinkle vanilla sugar over the cream-milk mixture. Add the vanilla bean. Scald over medium heat and do not stir.

Allow to steep for 1 hour, uncovered.

Into a large stainless-steel bowl, pour the egg yolks. Fill another, larger stainless-steel bowl with ice.

After the infusion period, remove the vanilla bean from the cream and slice it in half lengthwise. Scrape the seeds from the middle and into the cream. Toss in the scraped pod as well.

Rescald the cream, again not stirring. When the cream is just about to boil up and out of the pan, it is ready.

Quickly ladle a small amount of the hot cream into the yolks, whisking vigorously while doing so. Continue to ladle and whisk until about half the cream is in the yolk mixture. Pour the heated yolk mixture into the pan with the remaining cream-milk amalgam and immediately whisk vigorously. Scrape the bowl thoroughly into the pan.

Instantly, strain the crème anglaise through the chinois mousseline. Scrape the pan thoroughly into the chinois, forcing all of the custard through with a rubber spatula.

Chill on ice until completely cold, stirring frequently for the first 10 minutes of cooling

For the croque mademoiselle

Whisk cream, eggs, sugar, salt, and vanilla together in a bowl until completely combined.

In another bowl, fold almond extract and sugar into the marscapone.

Thinly spread 1 heaping tablespoon of the mascarpone mixture onto each slice of bread. Top half the slices with a thin layer of chopped chocolate. Sandwich the bread together so each one has a layer of chocolate.

Soak sandwiches in the custard for at least 1 hour per side.

Pan fry evenly on each side over medium heat until chocolate melts.

Dust with powdered sugar and serve with Crème Anglaise and whipped cream.

IGGY'S DOUGHBOYS

1151 Point Judith Rd., Narragansett, RI
iggysdoughboys.com, (401) 783-5608
(With a second location at 889 Oakland Beach Ave., Warwick)

Places like this make books like this a joy. I defy you to plunge your delicately battered clam fritter into a bowl of piping hot chowda (methinks the locals refer to this as "booning") and not ruminate about Norman Rockwell being inspired by doing the same. In fact, he probably did at Iggy's! Simply put, Iggy's is an institution; a Rhode Island mecca. Superlatives abound.

SOUNDTRACK SUGGESTION

"The Last Resort" by The Eagles

Don Henley's nod to Providence in the first half of this cut off *Hotel California* earns it its place in this book. It also captures the sense of desperation one feels when standing at the end of that long line that often plagues Iggy's.

Bemused by the very idea of a secret recipe, Iggy's goes the distance and puts their tried-and-true recipes directly on the side of the packages for both their fritter and doughboy mixes (available online or at either location). They're all about sharing the love.

And for those of you who don't know what a doughboy is: First of all, I am sorry for your upbringing. Second, it's like the fried dough you get at a carnival, only smaller and tastier, sans tomato sauce (probably known as "gravy" at Iggy's) in favor of cinnamon. You've had nothing else quite like one.

New England Clam Chowder
SERVES 4 TO 6

12 tablespoons (1½ sticks) unsalted butter, divided

2 cups chopped yellow onions (2 onions)

2 cups medium-diced celery (4 stalks)

4 cups peeled, medium-diced boiling potatoes (8 potatoes)

1½ teaspoons minced fresh basil leaves (½ teaspoon dried)

½ teaspoon black pepper

1 quart (4 cups) clam juice

½ cup all-purpose flour

2 cups half-and-half

2 cans Iggy's clams

Salt and pepper to taste

Melt 4 tablespoons (½ stick) of the butter in a large heavy-bottomed stockpot. Add onions and cook over medium-low heat for 10 minutes, or until translucent. Add celery, potatoes, basil, and pepper and sauté for 10 more minutes. Add clam juice, bring to a boil, and simmer, uncovered, until the vegetables are tender, about 20 minutes.

In a small pot, melt remaining 8 tablespoons butter and whisk in flour. Cook over very low heat for 3 minutes, stirring constantly. Whisk in a cup of the hot broth and then pour this mixture back into the cooked vegetables. Simmer for a few minutes until the broth is thickened.

Add half-and-half and clams and heat gently for a few minutes to cook the clams. Taste for salt and pepper. Serve hot.

Strawberry Doughboy

SERVES 6

1 package Iggy's Doughboy Mix

24 ounces vanilla ice cream

1 (24-ounce) container Cool Whip

4 cups fresh sliced strawberries

Cinnamon

Powdered sugar

Prepare half the doughboy mix according to package directions.

Place 2 ounces ice cream, ¼ cup strawberries, and 4 ounces cool whip on top of each doughboy. Sprinkle with cinnamon and sugar.

CRAZY BURGER CAFE & JUICE BAR

144 Boon St., Narragansett, RI
crazyburger.com, (401) 783-1810

As the name of the place implies, if you want to be health-conscious here, it ain't no big thang. One of Crazy Burger's craziest burgers—and a signature one to boot—is called the "Just Plain Nuts" burger (recipe below). It's a patty comprised of celery, carrots, zucchini, lentils, cashews, and walnuts. It has no right to taste as good as it does.

There is tofu in the house, but it's also BYOB. Basically, Crazy Burger is what you make it. Omelet additions/options run the gamut from gorgonzola cheese to black turtle beans. Oh, didn't I mention Crazy Burger serves breakfast?

In business since 1995, they've been booming on Boon ever since, in the heart of Narragansett Pier. The juice bar does those soaking up the sun just right, but the nachos, wings, and scrumptious sweet risotto-corn fritters can fill the most feverish bellies.

SOUNDTRACK SUGGESTION

"Cheeseburger in Paradise" by Jimmy Buffett

I mean, Buffett's ode to both cheeseburgers and paradise had to sneak into this book somewhere, so why not at a place that gets about as ambitious with the burger as one can possibly get, and on a pier that reeks of paradise?

Just Plain Nuts Burger
SERVES 4

4 cups water

2 cups dry mixed lentils

1 teaspoon canola oil

1 onion, diced small

2 stalks celery, diced fine

1 carrot, shredded

1 zucchini, grated

Salt and pepper to taste

1 cup cashews

1 cup walnuts

1 cup bread crumbs

4 hard rolls

1 cup ketchup

¼ cup mustard

Several basil leaves, washed and minced

Heat oven to 350°F.

Bring water to boil in sauce pot. Add lentils and simmer for 15 to 20 minutes. Drain and set aside.

Spread cashews and walnuts on cookie sheet and place in oven for 3 minutes until roasted.

In a sauté pan, heat canola oil over medium heat. Add onion, celery, and carrots until translucent, adding zucchini toward end of cooking time. Add salt and pepper to taste. Add breadcrumbs and transfer mix to medium bowl. Add lentils and combine.

Grind roasted cashews and walnuts and add to bowl. Portion mixture into burgers, adding more breadcrumbs if needed.

Grill a small piece to test for taste and consistency before grilling the rest of the burgers.

Place grilled burgers on rolls, garnish with ketchup or mustard as desired.

OCEAN MIST

895 A Matunuck Beach Rd., Wakefield, RI
oceanmist.net, (401) 782-3740

Okay, so you've gotta hang a right (or left) off Route One to reach this beachfront gem, but it's well worth it. A few minutes down the road and you'll feel like you're at a surfer bar in La Jolla. Especially if Dick Dale is jamming on stage, which is more than possible.

"We do a lot of reggae and a lot of blues here," said General Manager Katherine Gingras. "The original Sublime played on that stage!" How's that for California? But don't get it twisted, The Ocean Mist is all Ocean State. I took a selfie by a flyer for a sold-out Deer Tick show just days away; Rhode Island rockers with ridiculous indie cred.

The Tropical Breeze Salad comes with grilled chicken but you can substitute Yellow Fin Tuna Steak, both sit atop a bed of romaine, tomatoes, cukes, and Bermuda onion, with a colorful tropical fruit salsa rounding things out. Flavor-wise, it rivals the fruity concoction the locals refer to as a Kate Winslet, but the latter leaves ya buzzed.

SOUNDTRACK SUGGESTION

"Miserlou" by Dick Dale

That strumming, grab-your-board beat that kicks off "Miserlou" is probably the first thing that'll come to mind once you've gotten your exotic drink and you're staring off the back deck of the Ocean Mist, in the direction of Roy Carpenter's Beach and wishing you were born a surfer.

Tropical Breeze Chicken Salad
SERVES 4

FOR THE CHAMPAGNE VINAIGRETTE DRESSING

4 ounces champagne vinegar

1 teaspoon Dijon mustard

1 teaspoon honey

1 teaspoon shallots, finely minced

1 teaspoon garlic, chopped

Salt and pepper to taste

1 cup canola oil

FOR THE CHICKEN AND MARINADE

2 ounces of olive oil

2 ounces of apple cider vinegar

Juice of half a lemon

1 teaspoon garlic, chopped

½ teaspoon black pepper

½ teaspoon dry oregano

4 (6-ounce) chicken breasts, trimmed of excess fat

½ teaspoon your favorite jerk seasoning

FOR THE SALAD

1 head romaine lettuce trimmed, washed, and cut into bite-sized pieces

1 cucumber, peeled and sliced

20 cherry tomatoes

1 small red onion, thinly sliced

½ pineapple, peeled, cored and diced in ½-inch dice

For the dressing

In a blender, combine vinegar, mustard, honey, shallots, garlic, and salt and pepper, then slowly pour the oil in the small opening and blend the dressing until it is fully emulsified. Dressing may be stored in the refrigerator for up to a week.

For the chicken and marinade

Combine olive oil, apple cider vinegar, lemon juice, garlic, pepper, and oregano in a stainless steel or ceramic bowl, making sure the marinade is fully mixed.

Place the chicken breasts in the marinade, make sure they are well coated. Cover and place in refrigerator for 4 to 8 hours.

Remove chicken from marinade and sprinkle liberally with jerk seasoning.

On a hot grill, cook chicken breasts for 3 to 4 minutes per side, making sure the internal temperature reaches 165°F.

For the salad

Place lettuce on four chilled dinner plates. Ring with cucumbers and tomatoes and top with onions.

Slice chicken in to ¼-inch slices and place evenly on top of salads.

Finish with diced pineapples.

Drizzle with champagne dressing.

THE HAVERSHAM

336 Post Rd., Westerly, RI
thehaversham.net, (401) 322-1717

A hop, skip, and jump from Misquamicut Beach, The Haversham has history to spare, but looks decidedly into the future.

Chef Joe Dansereau is all about that fusion—no trouble. His Asian Seafood Paella (recipe below) is quite literally a treasure chest, loaded with enormous clams, equally sizable scallops, jumbo shrimp, and calamari (which he will enthusiastically tell you has just been decreed the state appetizer). He has a tattoo of a music note behind his ear, and his ear to the people. "They want to be motivated. Just like these clams. I've gotta motivate some of them to open up."

With Westerly having a strong Italian-American population, your chicken parm, linguini and clams, and caprese salad are present and accounted for, but Dansereau knows to expand upon that, too. Case in point: the grilled baseball steak, which is the perfect capper to a day on the beach.

Equal parts wedding facility and local tavern, The Haversham wants it both ways . . . and gets it, too.

Asian Seafood Paella

SERVES 2

6 cups chicken stock, divided

1 tablespoon plus 1 teaspoon curry powder, divided

2 cups uncooked rice

1 tablespoon garlic, chopped

½ cup onion, chopped

¼ cup red bell pepper, chopped

¼ cup green bell pepper, chopped

¼ cup chorizo, chopped

7 little neck clams

3 jumbo shrimp

1 cup calamari

½ pound sea scallops, chopped small

1 cup tomatoes, chopped with juice

Bring 6 cups chicken stock and 1 tablespoon curry powder to a boil. Add rice and return to boil. Cover with foil and simmer for 25 to 30 minutes. Set aside and make rest of dish.

Sauté garlic, onions, peppers, and chorizo until vegetables are translucent. Add remaining 2 cups broth, 1 teaspoon curry, and clams. Cover and simmer for 5 to 7 minutes.

Add shrimp, calamari, scallops and tomatoes and re-cover. Once clams open, remove and add rice. Simmer an additional 5 to 7 minutes until most liquid absorbed.

Dinner and a Movie

While I'm well aware that there are not a slew of road trippers who are inclined to both stop for a bite to eat and take in a movie at the same time, it is that elite group that I appeal to with this particular sidebar. Personally speaking, I'm all about grabbing a nice bite to eat after a day of driving and taking in some tried-and-true indie escapism.

Cable Car Cinema
204 S. Main St., Providence, RI
cablecarcinema.com

Cable Car is the cinemaphile's dream, man. Digging significantly deeper than your prototypical arthouse does, Cable introduces Ocean State residents to a ton of fare the rest of the country doesn't even know exists, furthering the theory that Rhode Island is an artist's playground. Sure, there's an accompanying cafe I can drone on about, but why would I when this is a movie house looking to tear stubs in half for flicks the film industry isn't even sure there is an audience for? Cable Car Cinema is the perfect ending to a day behind a wheel, and proof that independent cinema is alive and well.

THONG THAI RESTAURANT

15 Railroad Ave., Westerly, RI
thongrestaurant.com, (401) 348-0511

There is actually a lot of Thai to choose from between Connecticut and Maine, along Route One, but only one Thong, which is worth a pit stop if only for a pint of the pineapple fried rice.

Thong is a tiny, Rhode Island gem that will transport you to the streets of Thailand thanks to a menu that completely embraces its origins. The Tom Yum soup is delightful, while the soft-shell crab on a bed of steamed vegetables is a viable, undeniable alternative to the myriad Italian offerings swirling around your head.

The sticky rice with mango is seasonal fare at Thong, signaling the onset of summer in the Ocean State just as dramatically as a sudden 8:38 sunset.

SOUNDTRACK SUGGESTION

"The Sweet Escape" or "Hollaback Girl" by Gwen Stefani

The No Doubt songstress has a way of making wherever you are sound like it's a tropical locale via her solo material, and both of these songs are no exception. The former works in the sense that this restaurant is indeed an escape, while the latter is as playful as Stefani gets, and exactly how one feels here.

ANGELO'S CIVITA FARNESE

141 Atwells Ave., Providence, RI
angelosri.com, (401) 621-8171

This historic Federal Hill eatery was a favorite spot of my siblings and I when we would spend the summer in Rhode Island with our paternal grandparents. Approaching its 100th anniversary, simply put, Angelo's is more than an institution—it sparked a revolution.

In the early 1920s, Angelo's offered up Italian staples such as meatballs and gravy, macaroni and bean soup, and braciola. At the risk of alienating other wonderful restaurants spotlighted in this book, I have to say that the meatballs at Angelo's are second to none. They were the first I ever ordered at a restaurant, and the reason our grandmother stopped talking to all of us one weekend in 1976. (Hers were pretty spectacular, too.)

SOUNDTRACK SUGGESTION

"Brandy"
by Looking Glass

Angelo's is deserving of a '70s classic, and what better than this ode to sailors, or at the very least the woman many of them are ogling? It's a drinking song, but in the mellowest of ways.

My Americanized, McDonalds-frequenting self would plead with my grandparents to take me to the place where the meatballs are huge and served with french fries. I am certain there are Providence-born youth saying the same today.

Spring Fusilli

MAKES 4 SERVINGS

1 pound tri-color fusilli pasta

6 ounces oil

1½ tablespoons fresh garlic, chopped

3 cups fresh broccoli, chopped

2 cups fresh tomatoes, diced

1 pound chicken breast, fully cooked (grilled or sautéed) and diced

Salt and pepper to taste

Cook fusilli according to package directions.

While pasta is cooking, sauté oil and garlic in large sauté pan until garlic is translucent.

Add broccoli and cook until garlic is golden brown, tossing or stirring often.

Remove from heat and add tomatoes and cooked chicken breast.

Add salt and pepper to taste and stir well.

Divide pasta into four dishes and spoon sautéed vegetables equally over the four dishes

Mangia!

ASPARA RESTAURANT

716 Public St., Providence, RI
apsarari.com, (401) 785-1490

Aspara is so unique it's completely worth straying from Route One to this admittedly off the beaten path part of Providence. Yes, there is Chinese food, and yes, it is traditional Chinese food, but it is not what you find in so many Chinese restaurants across the country.

There is pretty much every stir fry imaginable on the sizable menu, with my particular favorite being the Tropical Delight. It's traditional stir fry, but prepared with tomatoes, pineapple, and green onion in a tangy brown sauce. The Chinese Delight stir fry is also delicious, as is the Kampoa, with the latter tossed with whole peanuts.

Beyond Chinese, Aspara comes at ya with Thai, Cambodian, and Vietnamese delicacies, specializing in rolls and salads to die for. A dramatic tagline might be "Don't try this at home," but the fact is we've got a recipe for you right here, for the Baby Shanghai salad dressing, aptly demonstrating that the folks at Aspara are down with you throwing down in the kitchen.

SOUNDTRACK SUGGESTION

"Every Little Thing She Does Is Magic" by The Police

Sting's tentative vocals at the outset of this early '80s anthem nail one's mood when first perusing the menu. And when the rousing, jubilant chorus kicks in it's like your order has just been placed before you!

Baby Shanghai

3 tablespoons oil

1 tablespoons garlic

2 tablespoons sugar

½ tablespoon salt

Pinch white pepper

10 slices fresh ginger

Combine all ingredients in a medium bowl and toss. Enjoy.

Dinner and a Movie

This place is a trip, both literally and figuratively. Do-able whether dining in Massachusetts or Rhode Island, it probably renders the "dinner" portion of this recurring section obsolete, as you can chow down right here in the theater, and not in a standard movie theater seat either.

Route One Cinema Pub
652 E. Washington St., North Attleboro, MA
cinema-pub.com
(508) 643-4618

Seriously, how does one not include a place called "Route One Cinema Pub" in a book called *Route One Food Run*? It's unthinkable! Plus, the joint is just so damn cool. They go for the occasional indie darling, but for the most part this is a place for mainstream cinema, replete with restaurant-style tables just feet away from an enormous, state-of-the-art screen. They do a Wednesday night date night (just $20) and—you guessed it—wine and beer is available too. A pub crawl that includes a flick? I'm in.

OLNEYVILLE NEW YORK SYSTEM RESTAURANT

18 Plainfield St., Providence, RI
olneyvillenewyorksystem.com, (401) 621-9500
(With a second location at 1012 Reservoir Ave., Cranston)

Hot damn, there's nothin' like a hot wiener. And you'd think they'd be enough. But here at Olneyville, they are bolstered by an unimaginably wonderful secret sauce. Bolstered and buoyed.

Then again, how secret could the sauce possibly be when you're provided with the recipe right on their website?

Lest you think the secret lay in the mustard or even the onion, the former is pretty much a mild deli mustard and the latter is a Spanish onion. In short, this thang is doable in your own backyard, folks.

But why would you deprive yourself of a visit?

Having recently celebrated 70 years serving greatness (can you tell I am a fan of the hotdog as a work of art?), Olneyville's story is truly one for the history books. Originally purveyors of sweets in Brooklyn (by way of Greece) in the 1930s, the Stevens family opted instead for the Ocean State and wieners.

Armed with the tagline "from lunch to late night" and a wiener that's halfway to a hero, Olneyville finds itself straddling the line between America's favorite luncheon snack and full-on gorge-yourself meal.

> ## SOUNDTRACK SUGGESTION
>
> "New York"
> by Alicia Keyes
>
> This head-boppin' groove-fest by Ms. Keyes is straight up necessity when it comes to a hot wiener from Olneyville, because it's all about feeling good. Plus, it is a "New York system" we're talking about after all.

Olneyville New York System Hot Wieners

SERVES 1

1 tablespoon soybean oil

1 Spanish or yellow onion, chopped

1 package Olneyville's Spice Mix

3 pounds fresh hamburger (not more than 80 percent lean)

1 beef, pork, and veal hot dog in natural casing

1 hot dog roll

1 teaspoon yellow bran mustard (mild deli mustard)

1 tablespoon special meat sauce,

1 tablespoon fresh chopped Spanish onion

½ teaspoon celery salt

For the sauce

Heat soybean oil in a 2-quart saucepan. Add onion and cook until brown.

Stir in the spice mix and ground beef. Cover and simmer for 1 hour.

For the weiners

The simplest way to do this is with a skillet with a lid. Place the hot dogs in the skillet, add 1 inch of water to the bottom and cover with a tight-fitting lid. Cook on high heat and add more water if you run out of steam before the hot dogs are fully cooked.

Steam the hot dog roll by placing it in a microwave for a few seconds. Place your cooked hot dog on the bun and top with mustard, 1 tablespoon of meat sauce, extra chopped onions, and celery salt to taste.

CHEF'S TIP

If the sauce becomes too dry, you can add additional soybean oil.

OGIE'S TRAILER PARK

1155 Westminster St., Providence RI
ogiestrailerpark.com, (401) 383-8200

While Ogie's sounds like a television series created by the same guy who gave us *Sons of Anarchy*, it bears much more of a resemblance to *Happy Days.* Except Arnold's didn't serve booze!

The decor endeavors to be an homage to the style and feel of mid-century America, from—according to the joint's Facebook profile—"that era between the birth of rock and the death of disco." But it's more cheeky than chic, and that's a good thing. The outdoor trailer park lounge is a pip, and the cocktails ample.

And the food . . . let's just say the eating's real good for trailer park living. Eggs Benedict on a biscuit with gravy, grilled watermelon, Brussels crisped and anointed with bacon and malt vinegar aioli for starters. Well, I'm not suggesting these be your appetizers. I'm suggesting you get here quick. In flip flops.

Ogie's French Canadian Tots

SERVES 1

1 cup of your favorite brand tater tots, cooked

2 slices slab bacon

1 egg

3 ounces of your favorite gravy, heated

goat cheese, to your liking

chopped scallions for garnish

Bake or fry tater tots according to package instructions until crispy.

In a pan, fry bacon until crispy. Allow to cool and then crumble. (This step can be omitted if you wish to make this dish vegetarian.)

In same pan, crack and fry the egg over easy.

Arrange tater tots on a platter. Place egg on top and drizzle with gravy.

Sprinkle with bacon crumbles, goat cheese, and scallions.

The runny egg, mixed with cheese and gravy over crispy tots gives this dish a hearty flavor and great textures. Enjoy!

JOE MARZILLI'S OLD CANTEEN ITALIAN RESTAURANT

120 Atwells Ave., Providence, RI
theoldcanteen.com, (401) 751-5544

Founded in 1956 by—you guessed it—Joe Marzilli, the Old Canteen sits comfortably in the center of Providence's Little Italy. When you hit the beginning of historic Federal Hill (my father's birthplace), look for the blinking sign to your left.

Just as quickly as your waiter will tell you the place is the oldest family-owned restaurant in all of Providence, he'll rattle off specials that are, moreover, the tried and true Italian fare of one's childhood. Beef braciola; veal Genovese; a half a chicken cut up, meat falling off the bone. My fave? The hunter-style veal and sausage, where the red sauce has simmered the meat "the rest of the way."

The appetizer must is the little neck zuppa, with little neck clams, netted just a few miles away in Rhode Island waters, popped open to perfection in a spicy red sauce. Wash it down with a Narragansett Shandy if it's summertime.

SOUNDTRACK SUGGESTION

"Suspicious Minds" by Elvis Presley

You'll feel like a king here, so why not cue some of the King up? And if you're going to, I say the rampantly revisited "Suspicious Minds" is the way to go, with the line "I'm caught in a trap" never resonating more gloriously.

Osso Bucco

SERVES 1

1- to 2-inch Osso Bucco (hind shank veal)

Salt and pepper to taste

1–2 tablespoons olive oil, for coating pan

¼ cup onion, diced

¼ cup celery, diced

¼ cup carrots, diced

3 cups tomato, diced

1 teaspoon oregano

1 teaspoon garlic powder

3 large bay leaves

Water to cover

Heat oven to 350°F.

Season osso bucco with salt and pepper.

Heat oil in a sauté pan over high heat. Sear veal on both sides, then bake in oven for 45 minutes. When done, place in baking dish.

Add onion, celery, carrots, and tomatoes. Sprinkle with oregano and garlic powder. Add bay leaves. Add enough water to the pan to just cover all. Cover with foil and continue baking for 2 hours.

Coffee Milk

I remember the day my father handed me a glass of ice cold coffee milk. Being the proud former Federal Hill resident he was, he beamed as he told me I wouldn't be able to get it anywhere else. He said the same of coffee-flavored ice cream.

This was some time around 1973 or so. It took years before I realized what he was truly saying; that his home state had their own weapon in the treat arsenal, something all their own. Whether it was true or not at the time (coffee-flavored ice cream was everywhere by the time Ben met Jerry), or something he thought to be true, I've no idea. I do know that as I grew older and began my own travels the mere mention of a coffee milk, from Hollywood, Florida, to Anaheim, California, generated the blankest of expressions.

The fact is, coffee milk became so popular in Rhode Island that in 1993 the Rhode Island state legislature voted it the official state drink. It's called a "cabinet" because its originator kept his blender in a kitchen cabinet.

What follows is an interview with David and Mary Sylvia, whose Morning Glory Old-Fashioned Coffee Syrup ships these days in Massachusetts and Rhode Island, and to Connecticut, Vermont, New Hampshire, and even New York.

VP: Not many people across the United States even know what coffee milk is! Why do you think that is, and why has it stayed Rhode Island's favorite treat for so long?

The Sylvias: We have often wondered this, and since we started selling our coffee syrup we have been educating potential customers to the staple of coffee milk. Given the coffee craze in the country and the fact that there is a Dunkin Donuts on almost every corner, one would think that what Rhode Islanders and south coast Massachusetts residents have known all these years would spread like wildfire. Most people are familiar with chocolate milk, so it's just a matter of substituting coffee syrup for chocolate syrup.

Coffee milk and coffee syrup is not only a Rhode Island favorite, but also a southeastern Massachusetts favorite. The first coffee syrup was actually produced by Silmo Packing Company of New Bedford, Massachusetts in 1932. We started our business in 2001, right around the time Silmo went out of business. The staying power of coffee milk in this region could be based on the large Portuguese and Italian populations that settled in this part of New England. Go farther north and the public has no idea what coffee milk is. In fact, when you ask for coffee milk outside of our area, you are brought a coffee and a cup of milk. Combine coffee syrup, milk, and coffee ice cream and you have yourself a coffee frappe or cabinet.

VP: What would you say to a non-coffee drinker who hasn't tried coffee milk in his/her life?

The Sylvias: We would say that you don't have to like hot coffee to enjoy coffee milk. Coffee milk does not have a bitter taste like coffee. When you make coffee milk with Morning Glory Old Fashioned Coffee Syrup it has a balanced blend of coffee flavor and sweetness. We would tell people to give it a try! You won't be disappointed. Lots of parents use coffee syrup to get their kids to drink milk. If the person is concerned about the caffeine content, we produce a decaffeinated version as well.

VP: How many restaurants do you ship Morning Glory to?

The Sylvias: We currently ship to several restaurants in our south coast area. We sell to Cork's Wine and Tapas bar in New Bedford where they have a signature drink, Neu Beige Coffee Milk, with whisky, double expresso vodka, and Morning Glory Coffee Syrup. We also sell to Bayside Restaurant and The Back Eddy in Westport; Grays Daily Grind in Adamsville, Rhode Island; Like No Udder (a mobile ice cream truck) in Providence; and Mermaid Farm on Martha's Vineyard where they make a coffee yogurt; and other ice cream establishments and eateries that are supplied through our distributor.

MASSACHUSETTS

CHARLIE'S SANDWICH SHOPPE

429 Columbus Ave., Boston, MA
charliessandwichshoppe.com, (617) 536-7669

Fast approaching 90 years of slinging hash, Charlie's is a true Boston haunt. In 1927 a seventeen-year-old went to work for Charlie in this place that rumor has it had no key, as it was open 24/7, rendering such a thing futile. In 1946 that very kid bought half ownership from Charlie himself for $1,000. The handwritten agreement still adorns the decidedly non-contemporary walls.

Repeatedly snapping up Best Breakfast awards throughout Boston, Charlie's remains a fixture today, its turkey hash and eggs being the hangover remedy of all hangover remedies, and the Cape Cod French toast ("buried in hot cranberry compote") a thing of freakin' beauty.

For lunch nosh on franks and beans or a meatloaf that rivals mom's. In fact, that's what Charlie has always sought to do: be a close second to your mom. If he bests her . . . well, so be it.

Charlie's Steak and Cheese

MAKES 1 SANDWICH

oil
salt
rib steak
onions
mushrooms
peppers
slices cheese (optional)
1 sub roll

Cut steak. For the primo steak 'n' cheese experience you want to slice it as thinly as possible, even getting to a point of chopping it.

Heat oil in a sauté pan over medium heat. Add onions, mushrooms, and peppers and salt to taste and sauté until onions are translucent.

In a separate pan, sauté steak until done to your liking.

Slice sub roll down the middle and toast in oven. Add cheese if desired and allow it to melt on roll. Add steak and top with onions, mushrooms, and peppers. Enjoy!

WARREN TAVERN

2 Pleasant St., Charlestown, MA
warrentavern.com, (617) 241-8142

One of the most historic taverns in the country, never mind just Massachusetts, The Warren Tavern actually counts George Washington and Paul Revere among its clientele. To its credit, the tavern does not shy away from such claims; heck, there's even a burger named the Paul Revere!

Another bountiful burger that shows up on the specials menu is the Sons of Liberty. This sucker comes topped with wild mushrooms, lettuce, tomato, and Gruyere cheese, with a swipe of garlic herb mayo on the bun serving as a bit of an exclamation point. It's superb. A thing of beauty.

For dinner, step it up and order the Tavern Special, in which chicken is accompanied by shrimp, prosciutto, sundried tomato, mushrooms, broccoli florets, and provolone, and it's all tossed in a garlic white wine sauce and served over linguine. Succulent, I tell ya.

The building is majestic, sitting on a Massachusetts corner and looking as if the *Cheers* bar could be right next door. It isn't, of course, but if you wanted everyone to shout out your name I am certain they would comply.

SOUNDTRACK SUGGESTION

"American Pie" by Don McLean

History meets history. McLean's time-honored classic may be about the day the music died, but it's also a harbinger of heft, of time having gone by, and the melancholy it would otherwise induce is bested by the inarguable spirit of the song.

Warren Tavern's Clam Chowder

MAKES APPROXIMATELY 1½ GALLONS OR 24 CUPS

1 pound butter

¼ cup onions, diced

3 ounces salt pork

1 pound flour

½ gallon clam stock

1¼ pounds potatoes, diced

4 pounds fresh clams, chopped

1 teaspoon garlic powder

1 teaspoon onion powder

1 teaspoon thyme

1 teaspoon cracked black pepper

1 teaspoon tabasco

1 teaspoon Worcestershire sauce

Melt butter in a large pot. Add onions and salt pork and sauté until onions are translucent.

Add flour and cook for 5 minutes.

Add clam stock and potatoes and cook until thickened.

Add chopped clams, garlic powder, onion powder, thyme, pepper, tabasco, and Worcestershire, and bring to a boil.

Remove salt pork, serve, and enjoy!

LA FAMIGLIA GIORGIO'S

112 Salem St., Boston, MA
lafamigliagiorgios.com, (617) 367-6711

The ambience! Tucked away neatly into Boston's North End, La Famiglia Giorgio's is the kind of place you take out-of-towners, if only for the selfies. It reeks of old New England, and also happens to reek of garlic at the same time.

SOUNDTRACK SUGGESTION

"Don't Wanna Miss a Thing" by Aerosmith

Boston's baddest boys need to find their way somewhere onto these pages, and where better than here? A family place for a band that's like family to so many of us, and a song about not wanting to miss a thing—which is surely how one would feel upon finding out this is where everyone went without me.

But this is not to say that La Famiglia is yet another in a moderately long line of Italian restaurants dotting this book's landscape—or Route One, for that matter. The Shrimp Gran Gala finds sizable black tiger shrimp plunged into Gran Gala orange liqueur, and finished with an orange wedge. The ravioli? Yeah, it's pumpkin.

But La Famiglia obviously goes the distance with the Italian cuisine, with every sauce you've ever tasted and/or heard of in your life on the menu. The Sorrento, with roasted peppers, onion, spinach, and canneloni beans in a white wine sauce is exceptionally good. You'll find every style of pasta and macaroni too.

Specializing in Roman cooking, this one is more destination than pit stop. Your stomach will thank you, and the kids will stop fighting. The stuff of reprieve.

MODERN PASTRY

257 Hanover St., Boston, MA
modernpastry.com, (617) 523-3783

Modern deserves more than a nod in our dessert section. This North End must-visit is worth so much more than that. Any meal must be capped by a cannoli and espresso at Modern, a Little Italy premier stop.

Many who wait in line to experience Modern go with traditional Italian baked goods, such as the pignoli cookies or sfogliatella, the latter as light and flaky as can be, despite more than a spoonful of a home-made yellow cream one bite away, but the American classics are where it's at.

SOUNDTRACK SUGGESTION

"Do You Believe in Magic"
by Lovin' Spoonful

Brimming with impishness and folk era fun, John Sebastian's presumably rhetorical question set to up-tempo music is as mercurial as Modern.

Boston Cream Cake is an obvious place to reside, but might I suggest the Elephant Ears. With cinnamon in every bite, the puffed pastry is baked to just the right point, golden brown, luxurious, and warm. Even the orneriest of travels will succumb.

Plus, the exterior, boasting early-twentieth-century charm, practically beckons.

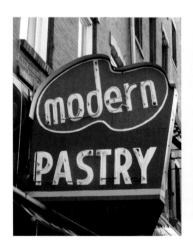

Dinner and a Movie

Literally nineteen minutes from Boston (and for road trippers that's basically a minute and a half), this independent movie theater is probably the best-looking of the lot, with its turn-of-the-century marquee and yesteryear trappings.

Coolidge Corner Theater
290 Harvard St., Brookline, MA
coolidge.org
(617) 734-2501

Considered New England's most successful independent, not-for-profit cinema, the Coolidge was built as a church in 1906. One can almost see the steeple in the pointed marquee. It was redesigned as an art deco movie palace in 1933, but even back then the mission was not to bring Abbott & Costello premieres. Its goal then was the same as its goal today: to bring contemporary art house and independent films. Yeah, they do the midnight cult screenings and even sing-alongs, but at its heart it's art.

KOWLOON RESTAURANT

948 Broadway, Saugus, MA
kowloonrestaurant.com, (781) 233-0077

Simply put, Kowloon is one of a kind. With its eye-popping exterior (it belongs in a sci-fi blockbuster) and equally indulgent menu, it's where you go to celebrate, and celebrate big. In fact, Kowloon would make a ton of sense being your final *Route One Food Run* stop.

SOUNDTRACK SUGGESTION

"Let's Get Rocked" by Def Leppard

Raise that Scorpion Bowl high and head bang along to the tasty guitar licks provided here, from this lead track off 1987's *Hysteria* CD, where a break from the day's chores/work is the central theme—as it is always at Kowloon's.

This is Chinese food with an attitude. You'll find basil-fried rice; General Gau chicken (as opposed to Tsao's), tossed with Kowloon's own hot ginger sauce; and Boston Maki, to name but a few outrageous items on the menu. For dessert you'll find Kowloon Fried ice cream or the mightiest of Mai Tais to wash your meal down.

Truth be told, many in the area head over for the Mai Tais alone, and to take in some live comedy. Yes, Kowloon has been many things to many people for a good long time now. Since 1950 it has been a draw. The kitschy Polynesian decor alone is worth making a visit. The second-to-none sushi menu and creative, colorful drink menu is what'll get ya to stay, though.

Chicken Wing Recipe

3 pounds chicken wings

3 tablespoons salt

2 tablespoons white sugar

6 tablespoons water

6 tablespoons soy sauce

1 tablespoon gin

¼ teaspoon ground ginger

1 quart vegetable oil, for frying

Combine the salt, sugar, water, soy sauce, gin, and ginger. Put mixture in a zipper style plastic bag, add chicken and marinate for 24 hours or as long as possible, turning bag frequently.

In a large skillet over medium high heat fry marinated chicken wings in hot oil until golden brown, about 8 minutes each side. Ready to serve!

Kowloon Lobster Sauce
MAKE 2 SERVINGS

⅓ pound ground pork

1½ cup water

2 tablespoons black bean sauce

1 tablespoon garlic, minced

1 teaspoon molasses (optional)

cornstarch, as needed for thickening

1 egg (optional)

salt to taste

Place the ground pork in a wok or large sauté pan. Stir fry until the pork is about ¾ cooked.

Add the water, black bean sauce, and molasses. Thicken to taste with the cornstarch. (Mix cornstarch with some water before adding to the pan.) While stirring, add the egg. (optional)

Continue stir frying as the egg cooks and the sauce heats through. Season to taste.

The cooking time depends on the type of pan and temperature of your stove top.

Serve with rice.

RED'S KITCHEN & TAVERN

131 Newbury St., Peabody MA
redskitchenandtavern.com, (978) 531-7337

After a rousing chorus of "Are we there yet?" you'll burst though Red's doors with wild abandon. It's a place the whole ramshackle family will love, and you'll still have a few bucks in your wallet when you leave.

SOUNDTRACK SUGGESTION

"With or Without You" by U2

Bono's aching vocals here, singing of the ultimate, heartbreaking decision to go on with or without someone, has happened to many Bostonians simply when it comes to going to this top spot.

Red's gives you pancakes a whole bunch of ways, mac and cheese a whole bunch of ways, and grilled cheese a whole bunch of ways. It's like they know your kids have been driving you crazy.

As for you, head to the Classics portion of the menu, where I defy you not to go for the American chop suey, where its hamburger front and center, seasoned to perfection and tossed with elbow macaroni, amongst other tasty vittles. Or the turkey dinner, which packs a Thanksgiving punch that, for many parents, is as close to a thank you as you are ever going to get.

GALLERIA UMBERTO

289 Hanover St., Boston, MA
facebook.com/pages/Umberto-Galleria/121498671193473,
(617) 227-5709

Talk about playing by your own rules, this place is cash only, and open until they sell out, so their hours are all over the place. You know what you get when you're dealing with a place like this? Outstanding food. Umberto walks the walk. Sans talk.

Most Bostonians will tell you it's all about the pizza here, and while they wouldn't necessarily be wrong, I am here to tell you that you must—*must*—try the arancini. It has just the right amount of crunch and is loaded with the zestiest of meat. While the rice balls are equally divine, this is a rice ball taken to the next level.

There is a throwback allure to the setting; call it retro chic. The only stainless steel you'll find here is the silverware. Most customers are fully prepared to eat their slice standing up, and I can't blame them. Also, when it comes to a slice, might I suggest going Sicilian? They're quite sizable, close to deep dish but not quite, and boast a perfect crust. Here, it's all about the bones.

SOUNDTRACK SUGGESTION

"Smooth Criminal" by Michael Jackson

That infectious bass-line, the King of Pop's indecipherable lyrics, the sense that shenanigans are afoot—such is an excursion to Umberto.

Notes for Making the Perfect Apizza

MAKES 1 PIE

1 dough

Sauce

Cheese

The dough you choose is up to you. Folks at Umberto tell us Sicilian-style pizza dough must raise unlike Napolitano dough, which is placed in the oven right after it's stretched.

With Sicilian, after you stretch the dough, you place it in a pan to rise. Temperature and humidity determines how long it will take to rise; in the summer, when it's 90°F, it can raise in an hour. Always cover the dough and place it where there is no draft so it doesn't dry out. An an oven (not on of course!) is an ideal spot.

Your sauce should depend on the season and availability of tomatoes. Whether you choose Italian ground and peeled tomatoes or California ground and peeled tomatoes, grind them and put on the dough.

Blend provolone, cheddar, and mozzarella cheese to your liking and evenly distribute it on your sauced dough.

Bake until bubbling and lightly brown and enjoy!

SANTA CRUZ DELI & CREAMERY

1007 Boston Post Rd. East, Marlborough, MA
santacruzdeliandcreamery.com, (508) 229-8500

Santa Cruz is the kind of deli where you really want a sandwich on the menu named after you. The Vinnie features salami, hot capocolla, sweet peppers, and provolone cheese on a multi-grain bread. The names of some of the sandwiches are as enjoyable as the sandwich itself.

Take for instance the Real McCoy (the tastiest of glazed maple hams loaded with onions, pickles, and barbecue sauce) or the Shark Bite (tuna salad nestled on provolone with banana peppers and "tsunami sauce," whatever that is).

SOUNDTRACK SUGGESTION

"Hold On"
by Wilson Phillips

You need the full California experience when it comes to this place, so a song from second generation Californians makes total sense, plus the lush vocal harmonizing can really help with digestion.

The sandwiches/subs come served on Boston's own Piantedosi bread, which dates back to 1916, while the decor is Art Deco fab. This mash-mash of time period, cross-pollination of generation, only adds to the spectacle.

Cap your visit with a cup or cone filled with extra creamy Santa Cruz soft serve and you'll practically hear those California waves crashing in the distance, the cones no different than a seashell, where the ocean sounds just moments away.

Santa Cruz Deli & Creamery's Famous Uncle Reuben Sandwich

MAKES 1 SANDWICH

1 pound sauerkraut

12 ounces lager beer

⅛ cup onion, thin-sliced

⅛ cup carrots, shredded

3 apple slices

Salt, pepper, and paprika to taste

2 slices Piantedosi marble rye bread

3 slices Boar's Head Swiss cheese (sliced thin)

5 ounces Boar's Head top round corned beef (sliced thin)

1-2 ounces Ken's Thousand Island dressing

1 ounce banana peppers

Combine all sauerkraut, beer, onion, carrots, and apple slices in a large sauce pot and simmer for 30 minutes to meld flavors and cook off the alcohol. Add salt and pepper to taste as you go.

When sauerkraut is close to being done, begin to prepare your sandwich. Place bread slices open faced, with cheese on one side and meat on the other. Toast until outside of bread is golden brown, cheese is melted, and meat is warm. Add sauerkraut, dressing, and banana peppers. Close sandwich and reheat for approximately 1 minute. (Take care not to add too much sauerkraut liquid or your sandwich will be soggy.)

Bean There, Done That

You're not thinking coffee, are you? Because, I'm talking baked beans. And not just any ol' baked beans—I'm talkin' Boston baked beans, and some rather kickass places to find a heaping, hot damn good time of a bowl. This is Boston-centric, no North Andover and the like. This is maple syrup woven in, or slabs of tenderized bacon, or, as is the case with one establishment that opened in 1826, simply slow-cooked, not unlike kettle-born chowda. Oh yeah, pretty much all these spots double as places where you can get your chowda fix, too.

Bar Louie
121 Brookline Ave., Boston
(617) 449-7010

Durgin-Park
340 N. Market St., Boston
(617) 227-2038

Bostonia Public House
131 State St., Boston
(617) 948-9800

Union Oyster House
41 Union St., Boston
(617) 227-2750

Boston Burger Company
37 Davis Square, Somerville,
(617) 440-7361
1105 Mass Ave., Cambridge,
(857) 242-3605
1100 Boylston St., Boston,
(857) 233-4560

Mike & Patty's
12 Church St., Boston
mikeandpattys.com
(617) 423-3447

Beantown Pub
100 Tremont St., Boston
(617) 426-0111

Cheers
84 Beacon St., Boston
(617) 227-9605

State Street Provisions
255 State St., Boston
(617) 863-8363

Sweet Cheeks Q
1381 Boylston St., Boston
(617) 266-1300

Red's Best Seafood at Boston
Public Market
37 Fish Pier St. W., Boston
(617) 428-0033

Atlantic Fish Company
761 Boylston St., Boston
(617) 267-4000

Neptune Oyster
63 Salem St., #1, Boston
(617) 742-3474

Sam Lagrassa's
44 Province St., Boston
samlagrassas.com
(617) 357-6861

And in the end, the food you make is equal to the bread you break.

MAINE

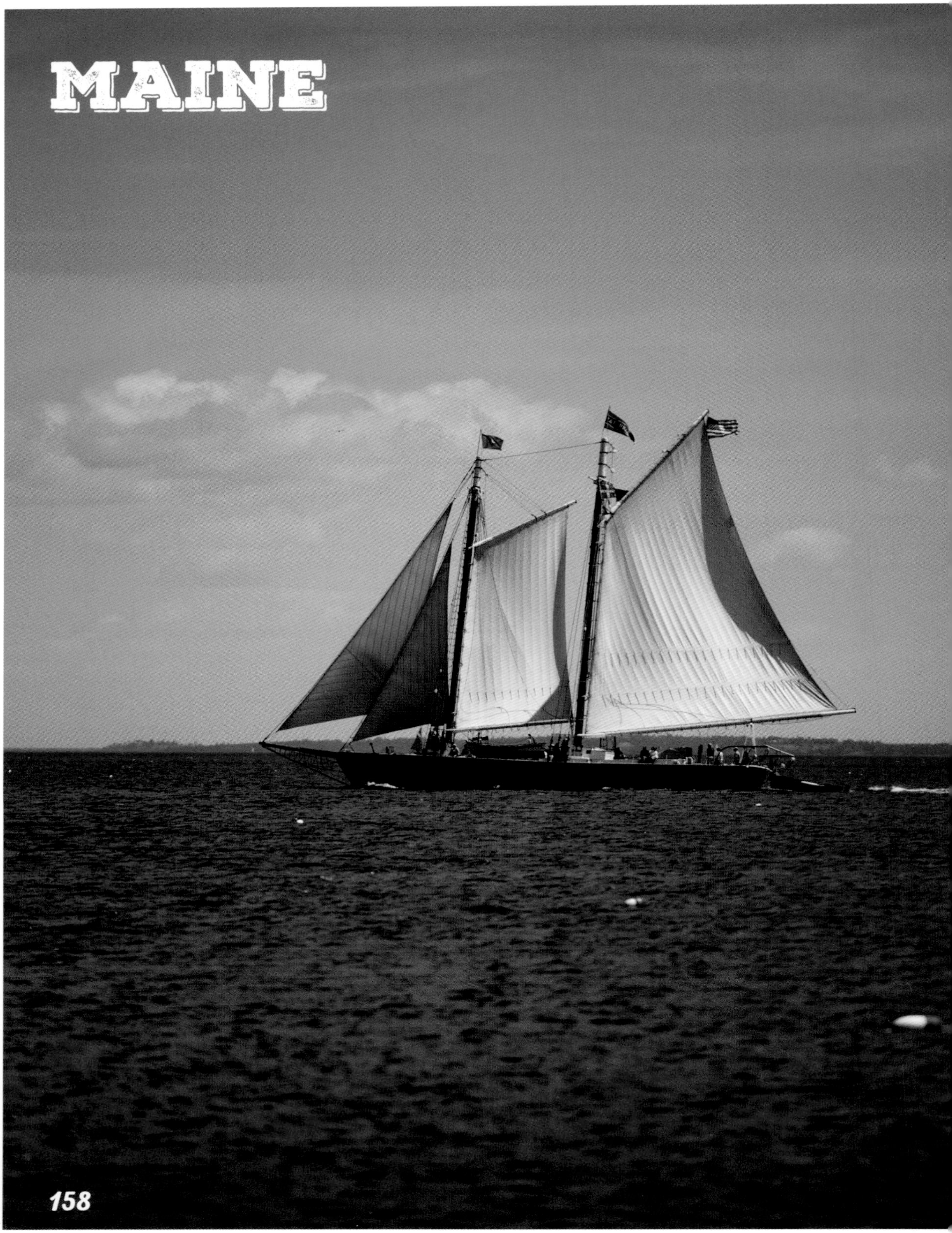

BOB'S CLAM HUT

315 US-1, Kittery ME
bobsclamhut.com, (207) 439-4233

Bob's opened in 1956, long before Kittery was besieged with shops for those trekking to Maine to indulge in local crafts and other impulse buying. Since that time a sort of mantra has emerged: "No visit to Maine is complete without going to Bob's." This is true.

SOUNDTRACK SUGGESTION

"Pulling Mussels" by Squeeze

The strum of the guitar, the buildup to the fun-filled chorus—what doesn't work here? It's a song about shell fish, for Pete's sake.

The lobster stew (recipe below), at once creamy and loaded with chunks of the most tender lobster meat, is the guiltiest of pleasures. Scratch that, as no one is guilty after gorging on a bowl.

No summer is complete without a clam or lobster roll. Here at Bob's they don't stop at those staples. Choose from succulent scallops, or shrimp, crabs, tuna, or oysters. If you want it on a roll, that's totally doable. Bob's has joined with another Maine force to be reckoned with; they now serve Rococo ice cream. The Kennebunkport favorite is a welcome addition, with their unique flavors such as goat cheese blackberry Chambord, or Maine whoopie pie. I once heard someone say that "Bob's is worth the pilgrimage," as if this humble eatery is the sole purpose for a journey to Steven King's playground. You know what? It is.

Lobster Stew

SERVES 6

2 tablespoons butter

1 cup finely Spanish onion, diced

2 cups lobster stock

2 cups ½-inch potatoes, cubed

1½ cups carrot, diced

1½ cups celery, diced

2 tablespoons fresh chives, chopped

1 tablespoon fresh parsley leaves, chopped

⅛ teaspoon cayenne pepper

Salt and freshly ground black pepper, to taste

2 cups (1 pint) heavy cream

1½ cups whole milk

1 tablespoon extra-virgin olive oil

1 pound fresh lobster meat, coarsely chopped into 2-inch pieces

⅓ cup sherry wine

In a stockpot, melt butter over low heat until bubbling. Do not burn.

Add onion, and caramelize over medium heat; do not let them brown.

Add lobster stock, potatoes, carrots, celery, chives, parsley, cayenne and 1½ cups water. Season with salt and pepper and bring to a boil over medium heat. Reduce heat and simmer until the potatoes are cooked (soft but hold up), about 15 minutes.

Add milk and cream and bring to a boil. Immediately turn down heat.

While the cream and milk are warming, add the olive oil to a sauté pan over medium-high heat. Once the oil is smoking, add lobster meat and season with salt and pepper to taste. When the lobster meat is warm, deglaze the pan with the sherry.

Add the warmed lobster meat and juices to cream mixture. Season with salt and pepper, if needed. Ladle into serving bowls and serve.

Dinner and a Movie

Maine has "Movies at the Museum" scant seconds from Route One, but I am loathe to begin recommending museums. This is road trippin' we're talking about after all! While the museum does boast quality deconstructions that, in a roundabout way, extoll the virtues of music to drive to, my goal here is to get you to a cool place at which to dine and, if you're so inclined, take in a gem of an independent film at a movie theater that oozes both history and charm. So, I present to you:

Eveningstar Cinema
149 Maine St., Brunswick, ME
eveningstarcinema.com
(207) 729-5486

The Eveningstar was born in 1979, basically the same year disco was born (and died). Its history is a novel in itself, storied to say the least, and ably documented at the website provided above. Suffice it to say that the Eveningstar rode the indie arthouse wave from its inception, if not playing a role in its creation, wavering somewhat in the '90s, only to return to its original incarnation. These are not the movies whose trailers will punctuate the SuperBowl; sequels and trilogies have no place here. The goal is something you've never quite seen before. And isn't that the goal of the road trip itself?

WASSES HOT DOGS

2 N. Main St., Rockland, ME
facebook.com/Wasses-332841964527, (207) 594-4347

Okay, so the prices here will cause a double take. Every dog they do comes in at under $2.00. Two bucks! Oops . . . the hotdog with bacon, chili, and cheese is $2.00 exactly. Look, the most expensive thing on the menu at Wasses is the bacon cheeseburger, coming in at a whopping $2.40. Simply put, you can't lose. If you're a Little League coach, you're crazy to go anywhere other than here after your team wins, and especially after they lose.

SOUNDTRACK SUGGESTION

"The Power of Love" by Huey Lewis & The News

This joint is such a throwback that a song associated with the whole throwback theme is key here, and this *Back to the Future* theme is . . . well, that theme.

The structure is a little white hut, harkening back to yesteryear, to simple pleasures and simpler times. Wasses is basically a time machine.

There are Wasses hot dog stands all over central coastal Maine and why wouldn't there be? Just as Del's lemonade stands used to dot Rhode Island's landscape, Wasses is a Maine staple.

From Wasses Hot Dogs of Maine: "In business for over 60 years, we're a Maine institution. I've owned Wasses the last 45 years. There are four locations. As far as a recipe is concerned, I don't really have one. I mean, I do and I don't. Our hot dogs are cooked in peanut oil. That's the secret. But there's more. The bun is steamed. And a Wasses should be served with a combination of mustard, relish, and FRIED onions. Furthermore, the onions MUST be cooked in the same oil the hot dog before it was."

HOME KITCHEN CAFE

650 Main St., Rockland, ME
homekitchencafe.com, (207) 596-2449

There is something very exciting about the Home Kitchen Cafe, where the very color of the walls is popping. The whole place is a Technicolor daydream. Exclamation marks at the end of every item on the menu wouldn't seem out of place somehow.

Serving breakfast and lunch, entrees such as the spicy Home stir-fry or pan-fried haddock are sourced with great ingredients and generous portions. Speaking of haddock, the haddock chowder is a must, a veritable meal unto itself. Chock full of potatoes, onions, and fish, it's spectacular. So is anything you can get topped with the kitchen's sausage gravy, which maybe single-handedly puts the comfort in comfort food (they were also kind enough to share the recipe!)

But Home hangs its hat on its breakfast. It wins awards year after year, and one need only look at the ambitious selections on the menu to see why. Might I recommend the Lady in the House. It's Home Kitchen's version of the Croque Madam, a grilled ham and swiss on their homemade bread, smothered in cheese sauce, a dash of Dijon, and topped with an egg. Or, sure, you can keep it simple. The crazier fare is really just gravy.

SOUNDTRACK SUGGESTION

"We Are Family"
by Sister Sledge

A song celebrating family seems apropos when it comes to a place named Home, does it not? And the guttural "yeah, yeah, yeah-yeah" can go around the table like the home fries will.

Hollandaise Sauce

½ pound butter

6 eggs, separated

⅛ teaspoon salt

2 dashes hot sauce

2 ounces fresh lemon juice

1 teaspoon of warm water

Melt butter in sauce pan or microwave to nearly boiling.

Place egg yolks in bowl of food processor and add salt, hot sauce, and lemon juice. Fit a food processor with cutting blade and begin to process, slowly adding in warm water.

Very slowly add the nearly boiling butter while food processor is running.

Let mix for 1 minute, seasoning to taste with more salt, lemon, or hot sauce.

If sauce is too thin, empty the hollandaise into a stainless steel bowl, place bowl on stove top burner on low heat and whisk continuously until sauce thickens just enough for you to see a swirl pattern with the whisk. Do not stop whisking until done.

Sausage Gravy
SERVES A LOT OF PEOPLE

6 cups milk

6 slabs bacon

5 pounds breakfast sausage

1 teaspoon dried sage

1 teaspoon dried thyme

Salt and pepper to taste

½ cup flour, plus 2 tablespoons

2 tablespoons butter

Heat milk in sauce pan over low heat.

While milk is heating, cook bacon in a heavy sauce pan until crisp, not burnt. Remove bacon, reserving the fat. Add breakfast sausage to the pot and cook slowly until crumbled and completely cooked. Add sage, thyme, salt, and pepper. Stir in ½ cup flour to sausage crumble and stir for 4 minutes to cook the flour. Add heated milk, stirring continuously until thickened.

Chop bacon and add to sausage gravy. If gravy is too thick, thin with milk. If not thick enough, make a roux of butter and flour and add a little at a time until desired consistency is achieved

Pour sausage gravy over biscuits or toast and enjoy!

MAINE DINER

2265 Post Rd., Wells, ME
mainediner.com, (207) 646-4441

An older gentleman I worked with for years, who called Maine home in the summer months, once described the Maine Diner to me as such: "There's always a line, and the customers sit on top of one another while eating." He may've even meant it as a slam, but I only heard, "You gotta go here. Everyone does." Presumably for a reason, no?

The short answer? Yes.

SOUNDTRACK SUGGESTION

"Our House"
by Madness

This second British invasion byproduct is as quirky as it is timeless, and the thumbing bass-line is the pitch-perfect backdrop to the hustle and bustle of a diner.

Known for their lobster pie, the Maine Diner is also a must for the seafood chowder and codfish cakes. Elsewhere, all three of these dishes could result in the type of game my sister and I would play as kids. We'd take respective bites, look at each other, and report if there was anything in said bite, as opposed to simply dough, crust, and/or broth. But not at the Maine Diner. Every single dish—every single bite—contains a chunk of seafood.

The Red Flannel Hash is breakfast the way it is intended to be served and enjoyed, and it wouldn't be a diner without a delicious meatloaf dish. Maine's comes on Mondays, with the nom de plume "Monday Meatloaf Madness." And, sure, the place gets so crowded its madness inside, too. Which is perfect.

Lobster Salad Plate

SERVES 1

5 ounce knuckle-claw meat, cooked

¾ ounce mayo

full bed of lettuce (4 oz)

green pepper slices

4 cucumber slices

4 cherry tomatoes

Mix lobster and mayo. Be sure to keep the mayo ratio low—you don't want it too mayonnaise-y, but you do need enough moisture. (Me, personally, I wanna taste lobster, not mayo!)

Spoon over lettuce and surround with peppers, cucumber, and tomatoes.

Makin' Whoopie:
An Ode to Maine's #1 Dessert

I sat down with Marcia Wiggins, the mastermind behind Cape Whoopie, a premier whoopie pie destination. Marcia is also a distributor of the spongy, creamy cakes. I had my own private tasting party; I fed her questions while she fed me pies. I consider it a win-win situation.

VP: How many Maine hotspots serve up your tasty cakes?

MW: Our biggest customer is Goldbely.com and they ship our product all over the country. In Portland, Maine, we are in Whole Foods Market, Captain Sam's Ice Cream, Gorgeous Gelato, Coastal Pharmacy, Fork Food Lab Tasting Room, Essex County Coop, and Go-Go Gifts in Boston's Logan Airport. There are a few others but that's a good list for now. On our web site we give all the locations.

VP: How would you describe a Whoopie Pie to the neophyte? To the Midwestern teen who thinks a Ring Ding is about as good as it gets?

MW: I would describe a Whoopie Pie to someone from the Midwest in this way: These treats are really much like a cake sandwich! We bake delightfully light, yet amazingly rich cakes that we fill with our gourmet cream filling. In Maine these cake sandwiches are the official state treat. Everyone loves them!

VP: Can someone make one in the comfort of his or her own home?

MW: There is a wide range of ways to make whoopies. Some are made with a filling that is Crisco and powdered sugar whipped together. We wanted to take that traditional treat and step it up a notch, so we make our own marshmallow from scratch and we add butter . . . then all the other flavorings to make our gourmet cream filling that absolutely melts in your mouth. Our cakes are very brownie-like with a richness that comes from the butter and cream cheese that we pack them with. We also felt that the amount of sugar in your typical whoopie pie was way too much. To us it was blocking all the other flavors! So, there's no powdered sugar in our pies, anywhere! You can taste the chocolate that we have flown in from the Netherlands and the vanilla we ship in from Madagascar. You'll just have to try these little babies and see why they have such a fun name, WHOOPIE!

VP: There is some steep whoopie pie competition around these parts. What makes Cape Whoopie stand out?

MW: We are small-batch whoopie pie bakers who love creativity in the kitchen. From the moment we tasted our first whoopie pie, we dreamed of endless flavor combinations with unique ingredients in a true gourmet indulgence. We believe in using only the freshest and finest ingredients, sourced from all over the world, remaining chemical and preservative free, and making everything from scratch.

Baking is our passion, and so is sharing our creations . . . if you love them like we do, you can find us at Whole Foods Market here in Portland or on Goldbely.com, DirectEats.com, and Amazon.com if you want to order for friends and have them shipped! We'd love to hear from you, and hope the quality and flavor of our pies exceed your every expectation.

We love to be creative and have seventy flavor combinations. We only keep ten flavors ready to go at a moment's notice, the rest we bring out, seasonally! Our regulars are:

Maine-Iac: chocolate cakes with our signature vanilla cream

Smart Blonde: vanilla cakes with our signature vanilla cream

Dark Strawberry: chocolate cakes with strawberry cream

Code Red: red velvet cakes with strawberry cream

Vanilla Latte: vanilla cakes with Khalua/espresso cream

Bromance: chocolate chip cakes with our signature vanilla cream

Marylou's Fav: chocolate cakes with peanut butter cream

Boston Strong: vanilla cakes with fudge and our signature vanilla cream

Britt's Fav: chocolate cakes with chocolate cream

Strawberry Shortcake: vanilla cakes with strawberry cream

In September we always unveil our fall flavors. This year, we're thinking about Drunken Punkin, which is a pumpkin spice cake with Amaretto cream; Granny's Gone Wild, with granny smith apple cakes and caramel cream; Crazy Nut Job, with walnut cakes and Frangelico cream and who knows what other flavor we're going to have? The sky is the limit, because we love creativity!

THE STEAKHOUSE

1205 Post Rd., Wells, ME
the-steakhouse.com, (207) 646-4200

I've learned that when a restaurateur keeps it simple with the name of his/her place, the mission statement is equally simple. Here it's all fairly straight-forward and obvious: a great cut of meat in a great atmosphere, with great service.

Check, check, and check.

Eye-popping beams adorn the dining room, catapulting the roof to extravagant heights, at once airy and cozy. Somehow.

The Steak Pot Pie is a hearty lumberjack delight, the filet mignon flavorful and sizable. Think chunk versus piece. Ditto the Steakhouse Chili, less an appetizer and more a full-on meal. The meat is tender, barely putting up a fight against the sizable Steakhouse forks.

The joint is cowboy chic, the food reasonably priced with king-sized portions, and the night woefully short.

Fresh-Made Lobster Stew

SERVES 1 (MULTIPLY RECIPE FOR THE AMOUNT OF PEOPLE)

3 ounces fresh Maine lobster meat

¾ ounces drawn butter

1 teaspoon lobster stew base

4 ounces half-and-half

Pinch freshly chopped parsley

Oyster crackers

In a copper-bottom, 1½-quart pot, place lobster meat, butter, and base. Sauté on high heat for about 45 seconds, then add half-and-half and continue to heat until bubbling but not scorched.

Transfer to your favorite bowl, add parsley, and serve with oyster crackers. Enjoy!

Filet Mignon with Baked Potato

1 9-ounce filet mignon, properly aged and well marbled

Salt and pepper to taste (or your favorite seasoning)

1 Idaho potato

Butter to your liking

Sour cream to your liking

Heat a gas grill to high. Sprinkle filet mignon with salt and pepper or your favorite seasoning and place on grill. Cook on high heat to sear in those juices, turning as needed, until meat reaches your desired temperature. We suggest medium rare.

Wrap Idaho potato in foil and place on grill. Cook to an internal temperature of 195°F.

Place filet mignon on plate along with potato. Spit potato in half and

top with butter and sour cream. Top filet with fresh ground black pepper and a scoop of your favorite butter. We like blue cheese crumbles and chopped garlic in ours. Enjoy!

Wild Maine Blueberry Crisp

MAKES 1 CRISP (OR 2 SMALL INDIVIDUAL CRISPS)

¼ cup rolled Quaker© oats

¼ cup brown sugar

¼ cup all-purpose flour

3 tablespoons butter, softened

1¼ cup wild Maine blueberries

1 tablespoon granulated sugar

Vanilla ice cream

Heat oven to 350°F.

In a small bowl, mix oats, brown sugar, flour, and softened butter. Combine well, making sure there are no big clumps.

In an oven-safe ramekin, layer blueberries and sugar, and top with crumb topping.

Place in oven for 20 minutes, cook longer if not golden brown and bubbling.

Serve with a generous scoop of ice cream.

WARREN'S LOBSTER HOUSE

11 Water St., Kittery, ME
lobsterhouse.com, (207) 439-1630

Warren's is a brand, man. And that stuff doesn't happen overnight. It's been seventy-five years, a gift shop has been erected, and still the tastiest lobsters this side of the tip of Maine can be found right here.

A dock and dine like no other (for the neophyte, this means you can pull your boat right on up to Warren's, drop anchor, and chow), it doubles as a spot where landlubbers get their tailgate on before picking out their prawn. What? You thought tailgating only happened before concerts?

With views of the Piscataqua River abounding, and one of the winner of the biggest restaurant sign award (seriously, the sign for this place looks like it belongs in front of a roller coaster), conversation will come as easy as a great meal. They have a lobster pound capable of holding twenty-eight tons of lobsters! The baked stuff sole is a good reason to eschew the delicacy in favor of something else, topped with a creamy hollandaise sauce and coming with the sixty-item salad bar. Yeah, that'd be the biggest salad bar on Earth pretty much.

Not for the faint of stomach.

> **"Come Sail Away" by Styx**
>
> Dennis DeYoung's soaring vocals, the tickling of the ivories, and the smashing of the cymbals once the band decides the verses are done is all a testament to stadium rock, boats being docked, and an epic meal being consumed.

Newburg Sauce

SERVES 2

2 tablespoons margarine

½ teaspoon paprika

2 tablespoons flour

1 cup clam juice

Pinch salt

¼ to ½ cup milk

4 tablespoons light cream

3 tablespoons sherry

Melt margarine over medium heat. Add paprika, and stir for 2 minutes. Add flour and stir for 2 to 3 minutes more, stirring constantly. Add clam juice and stir to thickening. Add ¼ cup milk, light cream, and sherry. Simmer for 5 minutes and, if needed, add remainder of milk. Serve over your favorite seafood.

INDEX

Additional photos: p. 138 Awirut Somsanguan/Shutterstock.com; p. 141 Charlie's Sandwich Shoppe © John Burns / www.flickr.com/photos/johncburns; p. 142 Charlie's Sandwich Shoppe © Brooks / www.flickr.com/photos/brooksbos; p. 147 Modern Pastry (left) © mfedore / www.flickr.com/photos/imfedore; (right) © Eris Delphi / www.flickr.com/photos/erisdelphi; Galleria Umberto: p. 154 © Ed Kopp / www.flickr.com/photos/edkopp4; p. 171: Darryl Brooks/ Shutterstock.com; pp. 5, 40, 58, 124, 130, 148, 163 Liu Zishan/Shutterstock.com

ACKNOWLEDGMENTS

I would like to thank the many, many restaurant owners and chefs I met while compiling this book, artists and visionaries all. Also, Jennifer Higham, whom I have had the pleasure of knowing since I was fifteen, and whose love of photography grew alongside my love of words. Your photographs are spectacular, and your odometer a sight to behold. Laurie Pennacchini, the unsung hero of this book, who I will sing about now; without her inimitable charm on the phone with each and every restaurant featured I doubt many would actually be featured herein. Without Laurie, therefore, there would be no book. My children Stella and Luke, who go on every adventure with me, ate some fine meals along the way this time around, and are a daily inspiration and miracle. Ann Nyberg, whom I affectionately refer to as "Connecticut Royalty," and who is not remotely aware that her reaching out to my editor on my behalf to start all this is not something everyone would do. And that editor! Amy Lyons, the most chill, laid-back editor with whom I have ever had the pleasure of working, equal parts creative and receptive: Thank you. Lastly, some folks who did a little homework for me in some of the areas I was less familiar: Michael Drago, Tony Bristol, and "Coach" George DeMaio. Anyone I have left out, I apologize—except for one or two who I did so intentionally.

ABOUT THE AUTHOR

Vinnie Penn is the host of the morning talk radio show The Vinnie Penn Project on New Haven's 960/WELI. Before that he spent close to ten years doing mornings on KC101. Route One Food Run is his third book, following a humor title and a zombie novel! His writing has been featured in Maxim, Cracked.com, MSN.com, Hit Parader, Circus, and Parent, plus you may have seen him on VH1's Best Week Ever, The Howard Stern Show, or Showbiz Tonight. He lives in Branford. Find him on Facebook (VinniePenn and TheRealVinniePenn), Twitter (@VinniePenn), and Instagram (@PennToTheV).

ABOUT THE PHOTOGRAPHER

Jennifer Higham is a premier wedding photographer in New England, and Route One Food Run is her first foray into the foodie world. She lives in East Haven, Connecticut with her two sons, RJ and Gianni. Find her at jhpct.com.